Farouk CHETOUAH

Planar antennas and miniaturization techniques

Farouk CHETOUAH

Planar antennas and miniaturization techniques

ScienciaScripts

Imprint

Any brand names and product names mentioned in this book are subject to trademark, brand or patent protection and are trademarks or registered trademarks of their respective holders. The use of brand names, product names, common names, trade names, product descriptions etc. even without a particular marking in this work is in no way to be construed to mean that such names may be regarded as unrestricted in respect of trademark and brand protection legislation and could thus be used by anyone.

Cover image: www.ingimage.com

This book is a translation from the original published under ISBN 978-620-6-72269-4.

Publisher:
Sciencia Scripts
is a trademark of
Dodo Books Indian Ocean Ltd. and OmniScriptum S.R.L publishing group

120 High Road, East Finchley, London, N2 9ED, United Kingdom
Str. Armeneasca 28/1, office 1, Chisinau MD-2012, Republic of Moldova, Europe
Printed at: see last page
ISBN: 978-620-8-35104-5

Contents

General introduction

Nowadays, the growing use and demand for various wireless communication devices in different fields, such as medicine, defence or aeronautics, motivates manufacturers to constantly develop new wireless communication systems. In this context, this technology has seen considerable progress in recent years in the field of microwave circuits. However, the antenna still occupies the largest volume in the communication chain, which implies an increase in overall size that makes it difficult to implement in small areas.

For several decades, antenna designers have been studying their miniaturisation with good performance in terms of gain or bandwidth. Several techniques have been proposed to reduce the size of antennas, such as the loading approach with a dielectric of very high permittivity, the use of inductive or capacitive ëlements, and short-circuit technology.

The ground plane defect structure (GDS) is another way of achieving miniaturisation and reducing the size of the antenna. In the scientific literature, there are a number of approaches using the DGS miniaturisation technique.

Despite their narrow bandwidths, microstrip antennas are still widely used in wireless communication networks because of their low manufacturing cost and simplicity of design. The aim of this work is to design and model planar antennae (microstrip antennae and dielectric resonator antennae), optimise their performance at UHF frequencies and manufacture these antennae for use in wireless communication systems. The aim of the proposed work is to study the potential for miniaturisation at microwave frequencies by combining electromagnetic radiation techniques with the characteristics of the chosen materials.

To carry out this work, two areas have been defined:

The first area concerns the choice of dielectric materials (with very high permittivities and low losses) for integration into planar antennae in order to achieve miniaturisation.

The second area concerns the manufacture and modelling of miniaturised planar antennas using the DGS (Defected Ground Structure) technique, as well as the development of the structure and improvement of the bandwidth.

The book is organised into the following five chapters:

The first chapter describes the basic concepts and principles of ylectromagnetism and plane waves.

General introduction

Chapter II includes дёпёгаШёз and ëlectrical and ëlectromagnëtic characteristics of planar antennas (microstrip antennas and dielectric resonator antennas).

In Chapter III, we will discuss the state of the art of the four miniaturisation techniques used in this research:

-      Modification of the site plan (DGS) ;
-      Loading with very high permittivity materials;
-      Integration of localised elements ;
-      Short circuit.

Chapter IV gives an overview of the properties of ultra-high dielectric permittivity resonators, together with a model and electromagnetic analysis of a miniature dielectric resonator antenna loaded with this material.

The final chapter is devoted to the fabrication of two miniaturised microstrip antennas using the DGS technique; the first antenna has a narrow bandwidth, while the second is broadband. The CST modelling software was used to analyse the antenna structures. In order to validate the numerical results, measurements were carried out. These results were satisfactory.

Chapter I

Properties of electromagnetic waves

1. Introduction

A uniform medium can be spëcifiedë by a small set of descriptive paramëtres whose behaviour of waves in such media can easily be described. Some of the key propriëtës of ëlectromagnëtic waves travelling in free space and other uniform media are prësentëes in this chapter. It ëestablishes the basic paramëtres and relationships that are used when consideringëre problems in antennas and propagation in later chapters.

2. Maxwell's equations [1]

2.1. Maxwell's equations

The existence of propagating ëlectromagnetic waves can be predicted as a direct consëquence of Maxwell's ëquations (1865). These ëquations spëcify the relationships between ëlectric field vector E and magnetic field vector H variations in time and space in a medium.

Maxwell's ëquations dëcrive all (classical) ëlectromagnëtic phënomënes:

$$\begin{cases} \nabla \times \vec{E} = -\dfrac{\partial \vec{B}}{\partial t} \\[2mm] \nabla \times \vec{H} = \vec{J} + \dfrac{\partial \vec{D}}{\partial t} \qquad \text{(Equations de Maxwell)} \\[2mm] \nabla \cdot D = \rho \\[2mm] \nabla \cdot B = 0 \end{cases} \qquad (1.1)$$

(Maxwell's equations)

The first ëquation is Faraday's law of induction, the second is Ampère's law modified by Maxwell to include the displacement current SD/dt, the third and fourth are Gauss's laws for ëlectric and magnetic fields.

The SD/5t displacement current term in Ampère's law is essential for predicting the existence of propagating electromagnetic waves. Equations (1.1) are in SI units. The quantities E and H are the electric and magnetic field strengths and are measured in units of [volt/m] and [ampere/m], respectively.

The quantities D and B are the electrical and magnetic flux densities and are expressed in units of [coulomb/m^2] and [weber/m^2], or [tesla].

D is also called the electrical displacement, and B, the nmignetic induction. The quantiles p and J are the volume charge density and electric current density (charge flux) of any external charge. They are measured in units of [coulomb/m^3] and [ampere/m].^{2}

The right-hand side of the fourth equation is zero because there are no magnetic monopole charges.

Charge and current densities p, J can be considered as the sources of electromagnetic fields. For wave propagation problems, these densities are localised in space; for example, they are limited to circulating on an antenna. The electric and magnetic fields generated are radiated far from these sources and can propagate great distances to receiving antennas.

An electric field is produced by a time-varying magnetic field. A magnetic field is produced by a time-varying electric field or by a current.

The qualitative mechanism by which Maxwell's equations give rise to propagating electromagnetic fields is shown in the figure below.

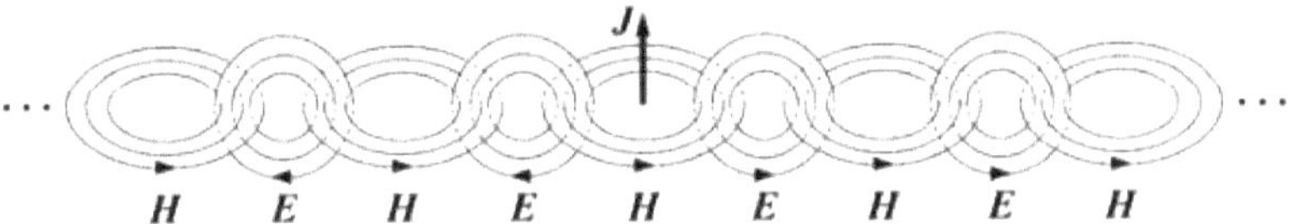

Figure I.1: Propagating electromagnetic fields.

For example, a time-varying current on an inear antenna generates a circulating and time-varying magnetic field H, which by Faraday's law generates a circulating electric field E, which by Ampere's law generates a magnetic field and so on. The cross-linked electric and magnetic fields propagate far from the current source.

2.2. Lorentz force

The force on a charge q moving with speed u in the presence of an electric and magnetic field E, B is called the Lorentz force and is given by:

$$F = q(E + v \times B) \quad \text{(Force de Lorentz)} \tag{1.2}$$

(Lorentz force)

Newton's equation of motion is (for non-relativistic speeds):

$$m\frac{dv}{dt} = F = q(E + v \times B) \tag{1.3}$$

Where; m is the mass of the load. The force F increases the kinetic energy of the charge at a rate equal to the rate of work done by the Lorentz force on the charge, i.e. u. F. In fact, the time derivative of the kinetic energy is:

$$W_{cin} = \frac{1}{2}mv \cdot v \quad \Rightarrow \quad \frac{dW_{cin}}{dt} = mv\frac{dv}{dt} = v\,F = qv \cdot E \tag{1.4}$$

We note that only the ëlectrical force contributes to the increase in kinetic energy; the magnetic force remains perpendicular to u, i.e.: $v \cdot (v \times B) = 0$.

The volume charge and current distributions p, J are also subject to forces in the presence of fields. The Lorentz force per unit volume acting on p, J is given by:

$$f = \rho E + J \times B \quad \text{(force de Lorentz par unité de volume)} \tag{1.5}$$

(Lorentz force per unit volume)

Or ; f is measured in units of [newton/m^3]. If J results from the movement of charges in the p distribution, then J = pu. In this case,

$$f = \rho(E + v \times B) \tag{1.6}$$

By analogy with equation (1.4), the quantity $v \cdot f = qv \cdot E = J \cdot E$ represents the power per unit volume of the forces acting on the moving charges, i.e. the power lost through the fields and converted into the charge's kinetic energy, or heat. It is expressed in [watts/m^3]. We will refer to it as:

$$\frac{dP_{perte}}{dV} = J \cdot E \quad \text{(pertes de puissance ohmiques par unité de volume)} \tag{1.7}$$

(ohmic power losses per unit volume)

dV

2.3. Constituent relationships

The electric and magnetic flux densities D, B are related to the field strengths E, H via so-called constitutive relations, the precise form of which depends on the material in which the fields exist. In a vacuum, they take their simplest form:

$$\begin{cases} D = \varepsilon_0 E \\ B = \mu_0 H \end{cases} \tag{1.8}$$

Or ; ε_0, po are the permittivity and permeability of the vacuum, with numerical values:

$$\begin{cases} \varepsilon_0 = 8.854 \times 10^{-12} \, farad \, / \, m \\ \mu_0 = 4\pi \times 10^{-7} \, henry \, / \, m \end{cases} \tag{1.9}$$

The units for so and po are the units of the D/E and B/H ratios, in other words:

$$\frac{coulomb \, / \, m^2}{volt \, / \, m} = \frac{coulomb}{volt.m} = \frac{farad}{m} \, , \quad \frac{weber \, / \, m^2}{ampere \, / \, m} = \frac{weber}{amper.m} = \frac{henry}{m}$$

From the two quantities so, po, we can define two other physical constants, namely the speed of light and the impedance characteristic of a vacuum:

$$c_0 = \frac{1}{\sqrt{\mu_0 \varepsilon_0}} = 3 \times 10^8 \, m \, / \, sec \, , \quad \eta_0 = \sqrt{\frac{\mu_0}{\varepsilon_0}} = 377 \, ohm \tag{1.10}$$

The next simplest form of the constitutive relations for a simple homogeneous isotropic dielectric and for magnetic materials is:

$$\begin{cases} D = \varepsilon E \\ B = \mu H \end{cases} \tag{1.11}$$

These are generally valid at low frequencies. The permittivity ε and the permeability μ are related to the electrical and magnetic susceptibilities of the material as follows:

$$\begin{cases} \varepsilon = \varepsilon_0 (1 + \chi) \\ \mu = \mu_0 (1 + \chi_m) \end{cases} \tag{1.12}$$

Susceptibilities χ, χ_m are measures of the electrical and magnetic polarisation properties of the material. For example, we have the electric flux density:

$$D = \varepsilon E = \varepsilon_0 (1 + \chi) E = \varepsilon_0 E + \varepsilon_0 \chi E = \varepsilon_0 E + P \tag{1.13}$$

Where; the quantity $P = \varepsilon_0 \chi E$ represents the dielectric polarisation of the material, i.e. the average electric dipole moment per unit volume.

In a magnetic material, we have:

$$B = \mu_0 (H + M) = \mu_0 (H + \chi_m H) = \mu_0 (1 + \chi_m) H = \mu H \tag{1.14}$$

Where: $M = \chi_m H$ is the magnetisation, i.e. the average magnetic moment per unit volume. The speed of light in the material and the characteristic impedance are:

$$c = \frac{1}{\sqrt{\mu\varepsilon}} \ , \quad \eta = \sqrt{\frac{\mu}{\varepsilon}} \tag{1.15}$$

The relative permittivity, permeability and refractive index of a material are defined by:

$$\varepsilon_r = \frac{\varepsilon}{\varepsilon_0} = 1 + \chi \ , \quad \mu_r = \frac{\mu}{\mu_0} = 1 + \chi_m \ , \quad n = \sqrt{\varepsilon_r \mu_r} \tag{1.16}$$

So that: $n^2 = \varepsilon_r \mu_r$.

Using the definition in equation (1.15), we can relate the speed of light and the impedance of the material to the corresponding vacuum values:

$$\begin{cases} c = \dfrac{1}{\sqrt{\mu\varepsilon}} = \dfrac{1}{\sqrt{\mu_0 \varepsilon_0 \varepsilon_r \mu_r}} = \dfrac{c_0}{\sqrt{\varepsilon_r \mu_r}} = \dfrac{c_0}{n} \\[4mm] \eta = \sqrt{\dfrac{\mu}{\varepsilon}} = \sqrt{\dfrac{\mu_0}{\varepsilon_0}} \sqrt{\dfrac{\mu_r}{\varepsilon_r}} = \eta_0 \sqrt{\dfrac{\mu_r}{\varepsilon_r}} = \eta_0 \dfrac{n}{\varepsilon_r} \end{cases} \tag{1.17}$$

For a non-magnetic material, we have $\mu=\mu_0$, ou, $\mu_r=1$, t the impedance simply becomes $\eta = \eta_0/n$.

In equations (1.1), the densities ρ, J represent the external or free charges and currents in a material medium. Induced polarisation P and magnetisation M can be explained in Maxwell's equations using the constitutive relations:

$$D = \varepsilon_0 E + P \ , \quad B = \mu_0 (H + M) \tag{1.18}$$

By inserting them into equations (1.1), for example, by writing:

$$\nabla \times B = \mu_0 \nabla \times (H + M) = \mu_0 (J + \dot{D} + \nabla \times M) = \mu_0 (\varepsilon_0 \dot{E} + J + \dot{P} + \nabla \times M),$$ Maxwell's

equations can be expressed in terms of the E and B fields:

$$\begin{cases} \nabla \times E = -\dfrac{\partial B}{\partial t} \\[4mm] \nabla \times B = \mu_0 \varepsilon_0 \dfrac{\partial E}{\partial t} + \mu_0 \left[J + \dfrac{\partial P}{\partial t} + \nabla \times M \right] \\[4mm] \nabla \cdot E = \dfrac{1}{\varepsilon_0} (\rho - \nabla \cdot P) \\[4mm] \nabla \cdot B = 0 \end{cases} \tag{1.19}$$

We identify the current and charge densities due to the polarisation of the material as:

$$J_{pol} = \frac{\partial P}{\partial t} \ , \quad \rho_{pol} = -\nabla \cdot P \qquad \text{(densités de polarisation)} \tag{1.20}$$

(polarisation densities)

Similarly, the quantity $J_{mag} = \nabla \times M$ can be identified as the magnetising current density (note that $\rho_{mag}=0$). The total current and charge densities are:

$$\begin{cases} J_{tot} = J + J_{pol} + J_{mag} = J + \dfrac{\partial P}{\partial t} + \nabla \times M \\[4mm] \rho_{tot} = \rho + \rho_{pol} = \rho - \nabla \cdot P \end{cases} \tag{1.21}$$

and can be considered as the sources of the fields in equation (1.19).

2.4. Negative index environment

Maxwell's equations do not exclude the possibilityë that one or both of the quantities £, p are negative. For example, plasmas below their plasma frequency, and metals down to optical frequencies, have $\varepsilon < 0$ et $\mu > 0$, with interesting applications such as surface plasmons.

Isotropic media with $\mu < 0$ et $\varepsilon > 0$ are more difficult to find [2], although examples of such media have been produced [3].

Negative-index media, also known as left-handed media, have r, ʟɪ which are simultaneously negative, $\varepsilon < 0$ et $\mu < 0$. Veselago [4] was the first to study their unusual electromagnetic properties, such as negative refractive index and inversion of Snel's law.

The new properties of these media and their potential applications have attracted a great deal of research interest. Examples of such media, known as 'metamaterials', have been constructed using periodic arrays of split-ring wires and resonators, [5] and by transmission line elements [6-8].

When $\varepsilon_r < 0$ et $\mu_r < 0$, the refractive index, $n^2 = \varepsilon_r \cdot \mu_r$, must be defined by the negative square root $n = -\sqrt{\varepsilon_r \cdot \mu_r}$ Because then $n < 0$ et $\mu_r < 0$ will imply that the characteristic impedance of the medium $\eta = \eta_0 \mu_r / n$ will be positive, which implies that the energy flow of a wave is in the same direction as the direction of propagation.

2.5. Boundary conditions

The boundary conditions for electromagnetic fields across material boundaries are given below:

$$\begin{cases} E_{1t} - E_{2t} = 0 \\ H_{1t} - H_{2t} = J_s \times \hat{n} \\ D_{1n} - D_{2n} = \rho_s \\ B_{1n} - B_{2n} = 0 \end{cases} \qquad (1.22)$$

Or ; *ni* is a unit vector normal to the boundary pointing from midpoint-2 to midpoint-1.

The quantities ps, ᴊs are external surface charges and surface current densities on the boundary surface and are measured in [coulombIm^2] and [amperelm^2].

In layman's terms, the tangential components of the E field are continuous across the interface; the difference in the tangential components of the H field is equal to the surface current density; the difference in the normal components of the D flux density is equal to the surface charge density; and the normal components of the B magnetic flux density are continuous.

The boundary condition ᴅn can also be written in a form that highlights the dëpendence on polarising surface charges:

$$(\varepsilon_0 E_{1n} + P_{1n}) - (\varepsilon_0 E_{2n} + P_{2n}) = \rho_s \;\Rightarrow\; \varepsilon_0(E_{1n} - E_{2n}) = \rho_s - P_{1n} + P_{2n} = \rho_{s,tot}$$

The total surface charge density will be $\rho_{s,tot} = \rho_s - \rho_{1s,pol} + \rho_{2s,pol}$, where the surface charge density of the polarisation charges accumulating on the surface of a dielectric is (f is the normal to the outside of the dielectric):

$$\rho_{S,pol} = P_n = \hat{n} \cdot P \tag{1.23}$$

The relative directions of the field vectors are shown in Figure I.2. Each vector can be decomposed as the sum of a part tangential to the surface and a part perpendicular to it, i.e.: $E = E_t + E_n$. Using the vector identity,

$$E = \hat{n} \times (E \times \hat{n}) + \hat{n}(\hat{n} \cdot E) = E_t + E_n \tag{1.24}$$

we identify these two parts as:

$$E_t = \hat{n} \times (E \times \hat{n}) \quad , \quad E_n = \hat{n}(\hat{n} \cdot E) = \hat{n} E_n$$

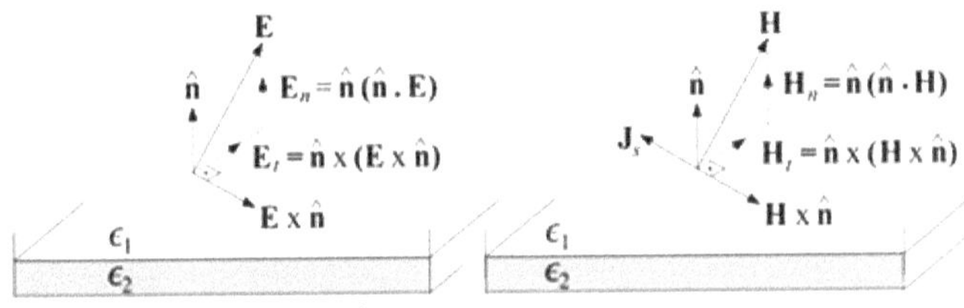

Figure I.2: Field indications at the boundary.

Using these results, we can write the first two boundary conditions in the following vector forms, where the secondième form is obtained by taking the cross product of the first with *ni* and noting that *Js* is purely tangential:

$$\begin{cases} \hat{n} \times (E_1 \times \hat{n}) - \hat{n} \times (E_2 \times \hat{n}) = 0 \\ \hat{n} \times (H_1 \times \hat{n}) - \hat{n} \times (H_2 \times \hat{n}) = J_s \times \hat{n} \end{cases} \text{ou,} \quad \begin{cases} \hat{n} \times (E_1 - E_2) = 0 \\ \hat{n} \times (H_1 - H_2) = J_s \end{cases} \tag{1.25}$$

The boundary conditions (1.22) can be derived from the integrated form of Maxwell's equations if we make some additional regularity assumptions about the fields at the interfaces.

In many interface problems, there are no surface charges or currents applied externally on the boundary. In such cases, the boundary conditions can be indicated as follows:

$$\begin{cases} E_{1t} = E_{2t} \\ H_{1t} = H_{2t} \\ D_{1n} = D_{2n} \\ B_{1n} = B_{2n} \end{cases} \qquad \text{(conditions aux limites sans source)} \tag{1.26}$$

(boundary conditions without source)

3. Plane wave properties [9]

There are many solutions to Maxwell's equations and all these solutions represent fields that could actually be produced in practice. However, they can all be represented as a sum of plane waves, which represent the simplest possible time-varying solution.

Figure I.3 shows a plane wave propagating parallel to the z axis at time t=0.

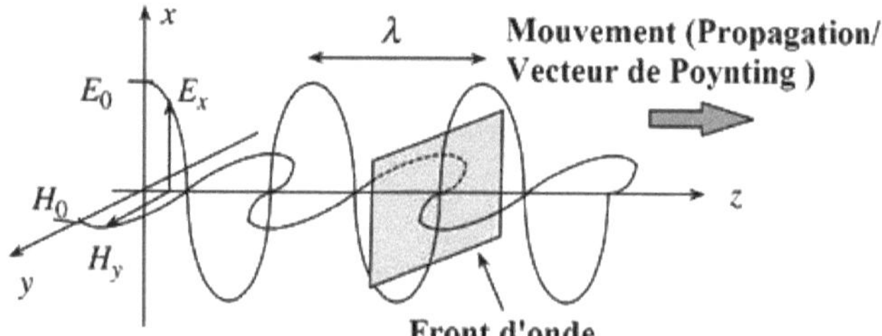

Figure I.3: A plane wave propagating in space at a given moment.

The electric and magnetic fields are perpendicular to each other and to the direction of propagation of the wave; the direction of propagation is along the z axis; the vector in this direction is the propagation vector or Poynting vector. The two fields are in phase at any time or in space. Their amplitude is constant in the xy-plane, and a surface of constant phase (a wavefront) forms a plane parallel to the xy-plane, hence the term plane waves.

The oscillating electric field produces a magnetic field, which in turn oscillates to create an electric field and so on, in accordance with Maxwell's curvature equations. This interaction between the two fields stores energy and therefore generates power along the Poynting vector. The variation, or modulation, of the wave's properties (amplitude, frequency or phase) then enables information to be conveyed in the wave between its source and destination, which is the central objective of a wireless communication system.

3.1. Field relationships

The ëlectric field can be written as:

$$E = E_0 \cos(\omega t - kz)\,\hat{x} \tag{1.27}$$

Where; Eo is the amplitude of the field [V m^{-1}], w=2rcf is the angular frequency in radians for a frequency f [Hz], t is the elapsed time [s], k is the wave number [m^{-1}], z is the distance along the z axis (m) and $_x$ is a unit vector in the positive x direction.

The wavenumber represents the rate of change of the phase of the field with distance, i.e. the phase of the wave changes by kr radians over a distance of r metres. The distance over which the phase of the wave changes by 2л radians is the wavelength. As well as:

$$(1.28)\, k = \frac{2\pi}{\lambda}$$

Similarly, the magnetic field vector H can be written as:

$$H = H_0 \cos(\omega t - kz)\,\hat{y} \tag{1.29}$$

Where; Ho is the amplitude of the magnetic field and $_y$ is a unit vector in the positive y direction.

In both equations (1.27) and (1.29), it has been assumed that the medium in which the wave travels is lossless, so that the amplitude of the wave remains constant with distance. Note that the wave varies sinusoidally in both time and distance.

It is often convenient to represent the phase and amplitude of the wave using complex quantities, so equations (1.27) and (1.29) become:

$$E = E_0\, e^{j(\omega t - kz)\hat{x}} \tag{1.30}$$

and

$$H = H_0\, e^{j(\omega t - kz)\hat{y}} \tag{1.31}$$

The real quantities can then be recovered by taking the real parts of equations (1.3o) and (1.31).

3.2. Wave impedance

Equations (1.27) and (1.29) satisfy Maxwell's equations, provided that the ratio of field amplitudes is constant for a given medium,

$$\frac{|E|}{|H|} = \frac{E_x}{E_y} = \frac{E_0}{H_0} = \sqrt{\frac{\mu}{\varepsilon}} = Z \tag{1.32}$$

Or ; Z is called the impëdance of the wave and has units of ohms.

In free space, $\mu_r = \varepsilon_r = 1$ and the impëdance of the wave becomes:

$$Z = \sqrt{\frac{\mu_0}{\varepsilon_0}} \approx \sqrt{4\pi \times 10^{-7} \times \frac{36\pi}{10^{-9}}} = 120\pi \approx 377\Omega \tag{1.33}$$

However, in free space or any uniform medium, it is sufficient to specify a single field strength with Z to specify the total field for a plane wave.

3.3. Phase speed

The speed of a constant phase point on the wave, the phase speed u at which the wavefronts advance in the direction S, is given by:

$$\upsilon = \frac{\omega}{k} = \frac{1}{\sqrt{\mu\varepsilon}} \tag{1.34}$$

Hence the wavelength is given by:

$$\lambda = \frac{\upsilon}{f} \tag{1.35}$$

In free space, the phase velocity becomes:

$$\upsilon = c = \frac{1}{\sqrt{\mu_0\varepsilon_0}} \approx 3\times 10^{8}\,ms^{-1} \tag{1.36}$$

Note that light is an example of an electromagnetic wave, i.e. c: the speed of light in free space.

3.4. Lossy environments

So far, only lossless media have been considered. When the medium has significant conductivity, the amplitude of the wave decreases with the distance travelled through the medium as energy is removed from the wave and converted to heat, so equations (1.30) and (1.31) are then replaced by:

$$E = E_0\, e^{[j(\omega t - kz) - \alpha z]\hat{x}} \tag{1.37}$$

And

$$H = H_0\, e^{[j(\omega t - kz) - \alpha z]\hat{y}} \tag{1.38}$$

The constant a is known as the attenuation constant, with units of per metre [m^{-1}], which depends on the permeability and permittivity of the medium, the frequency of the wave and the conductivity of the medium, σ, , measured by Siemens per metre or per ohm metre. [Ωm]$^{-1}$.

The set o, µ and e are known as the constituent parameters of the medium.

Consequently, the strength of the electric and magnetic field decreases exponentially as the wave moves through the medium, as shown in Figure I.4.

The distance travelled by the wave before its field reduces to e^{-1} =0.368=36.8 % of its initial value is its skin depth δ, which is given by:

$$\delta = \frac{1}{\alpha}$$
(1.39)

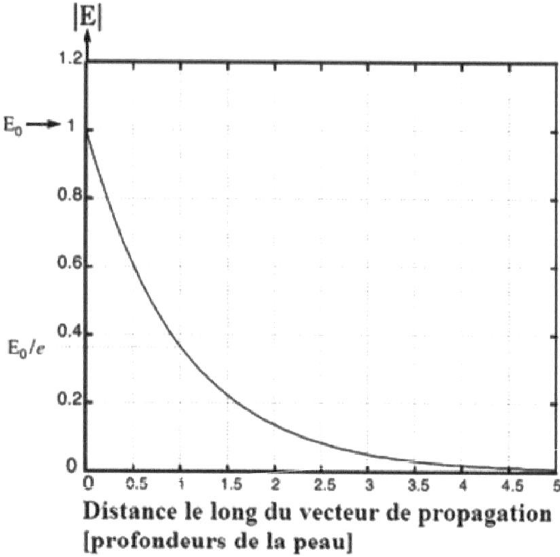

Figure I.4: Field attenuation in a lossy medium.

Thus, the amplitude of the intensityë of the electric field ë at a point z with respect to its value at z=0 is thereforeK'e by:

$$E(z) = E(0)e^{-z/\delta}$$
(1.40)

Table 1.1 gives expressions for a and k that apply in lossless and lossy media. Note that the expressions can be simplified according to the relative values of o and ws. If o dominates, the matëriau is a good conductor; if o is very small, the matëriau is a good insulator or diëlectrical.

Table I.1: Attenuation constant, wavenumber, wave impëdance, wavelength and phase velocity for plane waves in lossy media (from [10])

$n = ck/\omega$ dans tous les cas	Expression exacte	Bon diélectrique (isolant) $(\sigma/\omega\varepsilon)^2 \ll 1$	Bon conducteur $(\sigma/\omega\varepsilon)^2 \gg 1$
Constante d'atténuation α [m^{-1}]	$\omega\sqrt{\dfrac{\mu\varepsilon}{2}\left[\sqrt{1+\left(\dfrac{\sigma}{\omega\varepsilon}\right)^2}-1\right]}$	$\approx \dfrac{\sigma}{2}\sqrt{\dfrac{\mu}{\varepsilon}}$	$\approx \sqrt{\dfrac{\omega\mu\sigma}{2}}$
nombre d'onde k [m^{-1}]	$\omega\sqrt{\dfrac{\mu\varepsilon}{2}\left[\sqrt{1+\left(\dfrac{\sigma}{\omega\varepsilon}\right)^2}+1\right]}$	$\approx \omega\sqrt{\mu\varepsilon}$	$\approx \sqrt{\dfrac{\omega\mu\sigma}{2}}$
Impédance d'onde Z [Ω]	$\sqrt{\dfrac{j\omega\mu}{\sigma+j\omega\varepsilon}}$	$\approx \sqrt{\dfrac{\mu}{\varepsilon}}$	$\approx \sqrt{\dfrac{\omega\mu}{2\sigma}}(1+j)$
Longueur d'onde λ [m]	$\dfrac{2\pi}{k}$	$\approx \dfrac{2\pi}{\omega\sqrt{\mu\varepsilon}}$	$\approx 2\pi\sqrt{\dfrac{2}{\omega\mu\sigma}}$
Vitesse de phase v [ms^{-1}]	$\dfrac{\omega}{k}$	$\approx \dfrac{1}{\sqrt{\mu\varepsilon}}$	$\approx \sqrt{\dfrac{2\omega}{\mu\sigma}}$

4. Polarisation [9]

4.1. Polarisation states

The alignment of the electric field vector of a plane wave with respect to the direction of propagation defines the polarisation of the wave. In Figure I.3, the electric field is parallel to the x-axis, so this wave is x-polarised. This wave could be generated by a straight-wire antenna parallel to the x-axis. An entirely separate y-polarisée plane wave could be generated with the same direction of propagation and recovered indëpendently of the other wave using pairs of perpendicularly polarised demission and reception antennas. This principle is sometimes ^xИвë in satellite communications to provide two indëpendent communication channels on the same satellite link. If the wave is gënërëed by a vertical wire antenna (horizontal H field), then the wave is said to be vertically polarisëed; a wire antenna parallel to the ground plane (horizontal E field) gënëre mainly horizontally polarisëed waves.

The waves dëcribed so far have ël.ë polarisëes linearly, since the ëlectric field vector has a single direction along the axis of propagation. If two plane waves of equal amplitude and orthogonal polarisation are combined with a phase difference of 90°, the resulting wave will be circularly polarised (CP), in that the motion of the ëlectric field vector will describe a circle centred on the propagation vector. The field vector will rotate 360° for each wavelength travelled. Circularly polarised waves are most commonly used in satellite communications, since they can be generated and received using antennas that are oriented in any direction around their axis without loss of power. They can be managed in either right-hand circular polarisation (RHCP) or left-hand circular polarisation (LHCP); RHCP describes a wave with the electric field vector rotating clockwise looking in the direction of propagation.

In the most gënëral case, the components of 1 wave could be of unequal amplitudes or of a phase angle other than 90°. The result is an elliptically polarised wave, where the electric field vector still rotates at the same speed but varies in amplitude with time, describing an ellipse. In this case, the wave is characterised by the ratio between the maximum and minimum values of the instantaneous electric field, called the axial ratio, AR,

$$AR = \frac{E_{maj}}{E_{min}} \qquad (1.41)$$

AR is defined as positive for left polarisation and negative for right polarisation. These different polarisation states are illustrated in Figure I.5.

Direction of propagation (z-axis) is out of the page

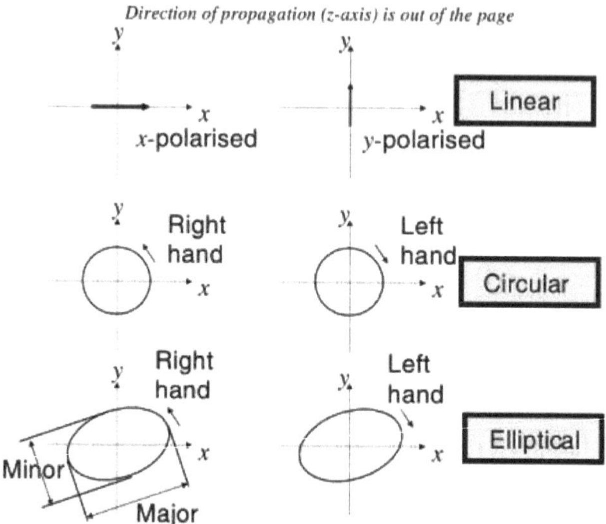

Figure I.5: Possible polarisation states for a z-directed plane wave.

4.2. Mathematical representation of polarisation

All the polarisation states shown in Figure I.5 can be represented by an electric field vector E composed of linearly polarised plane waves x and y with amplitudes E_x and E_y,

$$E = E_x \hat{x} + E_y \hat{y} \qquad (1.42)$$

Table I.2: Relative electric field values for the polarisation states shown in Figure I.5.

Polarisation state	E_x	E_y
Linear x	$E_0 / \sqrt{2}$	0
Linear y	0	$E_0 / \sqrt{2}$
Right-hand circular	$-E_0 / \sqrt{2}$	$jE_0 / \sqrt{2}$
Left-hand circular	$E_0 / \sqrt{2}$	$jE_0 / \sqrt{2}$
Straight elliptical trainer	$-aE_0 / \sqrt{2}$	$jE_0 / \sqrt{2}$
Elliptical left	$aE_0 / \sqrt{2}$	$jE_0 / \sqrt{2}$

The relative values of E_x and E_y for the six polarisation states in Figure I.5 are given in Table I.2, assuming that the maximum wave amplitude is E0 in all cases and that the complex constant a depends on the axial ratio. The axial ratio is given in terms of E_x and E_y as follows [11]:

$$AR = \left[\frac{1 + \left| \frac{E_y}{E_x} \cos\left[\arg(E_y) - \arg(E_x) \right] \right|^2}{\left| \frac{E_y}{E_x} \sin\left[\arg(E_y) - \arg(E_x) \right] \right|} \right]^{\pm 1} \qquad (1.43)$$

The liquidation exponent (1.43) is chosen such that $. AR \geq 1$.

5. Conclusion

The propagation of waves in uniform media can be described by considering the properties of plane waves, whose interactions with the medium are entirely specified by their frequency and polarisation and by the constituent parameters of the medium. Not all waves are plane waves, but all waves can be described by a sum of plane waves with an appropriate amplitude, phase, polarisation and Poynting vector.

Bibliography of Chapter I

[1]   consulted: http://www.ece.rutgers.edu/~orfanidi/ewa/ewa-1up.pdf
Sophocles J. Orfanidis, *Electromagnetic Waves and Antennas*, Rutgers University, 2016.

[2]   L. D. Landau, E. M. Lifshitz, and L. P. Pitaevskii, *Electrodynamics of Continuous Media*, 2/e, Elsevier Science, Burlington, MA, 1985.

[3]   J. B. Pendry, "*Magnetism from Conductors and Enhanced Nonlinear Phenomena*," IEEE Trans. Microwave Theory Tech, 47, 2075 (1999).

[4]   V. G. Veselago, "*The Electrodynamics of Substances with Simultaneously Negative Values of e and11*," Sov. Phys. Uspekhi, 10, 509 (1968).

[5]   D. R. Smith, et al, "*Composite Medium with Simultaneously Negative Permeability and Permittivity*," Phys. Rev. 84, 4184 (2000).

[6]   A. Grbic and G. V. Eleftheriades, "*Subwavelength Focusing Using a Negative-Refractive-Index Transmission Line Lens,* "IEEE Ant. Wireless Prop. Lett. 2, 186 (2003).

[7]   A. Grbic and G. V. Eleftheriades, "*Overcoming the Diffraction Limit with a Planar LeftHanded Transmission-Line Lens,* "Phys. Rev. Lett. 92, 117403 (2004).

[8]   C. Caloz and T. Itoh, "*Transmission line approach of left-handed (LH) materials and microstrip implementation of an artificial LH transmission line*," IEEE Trans. Antennas Propagat, 52, 1159 (2004).

[9]   Simon R. Saunders, Alejandro Arago'n-Zavala, *'Antennas And Propagation For Wireless Communication Systems'*, Second Edition, John Wiley & Sons Ltd, 2007.

[10]   C. A. Balanis, *Advanced engineering electromagnetics*, John Wiley & Sons, Inc, New York,

[11]   K. Siwiak, *Radiowave propagation and antennas for personal communications*, 2nd edn, Artech House, Norwood MA, ISBN 0-89006-975-1, 1998.

General information on planar antennas

1. Introduction

All antennas comprising radiating ëlements (with a flat or curved surface or variations thereof) and at least one feed are referred to, in general, as "planar antennas". [1]

In this first chapter, we will define two planar antennas: a microstrip antenna and a diëlectric resonator antenna. We will describe their main characteristics (electrical and electromagnetic) as well as their advantages and disadvantages, and will conclude with a comparison of the two technologies and their fields of application.

2. Antenna definition

An antenna can be used for transmitting and receiving electromagnetic waves, and is a device used to facilitate the transfer of energy between a transmission line and free space and vice versa (see figure II.1).

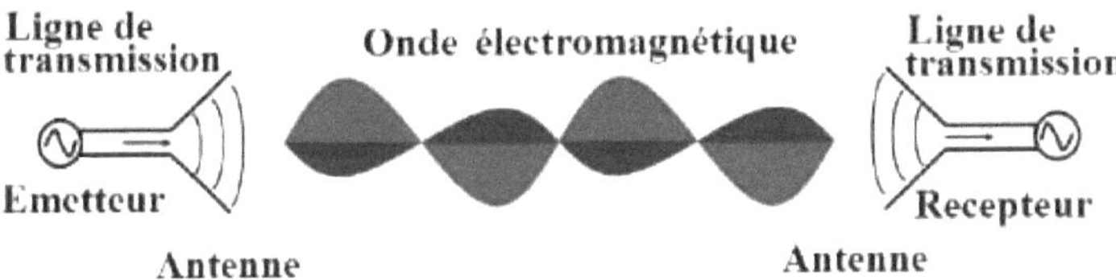

Figure II. 1: A typical radio system.

The IEEE standard definitions of antenna terms (IEEE Standard 145 -1983) [2] define an antenna as *"a part of a transmitting or receiving system designed to radiate or receive electromagnetic waves"*.

3. Antenna classifications [3]

Since the dawn of radio communications over 100 years ago, thousands of antennas have been ë!ë dëveloppers and ëtudiëed. They can be catëgorisëes by different crores:

- In terms of bandwidth, antennas can be classified as narrowband or broadband;
- From a polarisation point of view, they can be classified as linear, circular or elliptical polarisation antennas;
- From a resonance point of view, they can be organised in the form of resonant antennas (standing waves) or travelling waves;
- Depending on the number of elements, they can be grouped together in the form of single-element antennas or antenna arrays;
- From a construction point of view, they can be catëgorisëed into solid, liquid and gas antennas. Solid antennas refer to those made of conductive matërials (such as dipoles, loops and horns), diëlectrical matërials (such as DRAs) or a combination of the two (such as patch antennas). Liquid antennas are mainly made up of liquid types (1 plasma antenna uses a ëlëment of plasma as a conductive medium for the RF signal to be ravonne).

4. Terminology and electrical and electromagnetic characteristics of antennas

To describe the performance of an antenna, various parameters need to be defined.

4.1. Electrical characteristics [4, 5]

4.1.1. Reflection coefficient

This is the s_{11} parameter of the dispersion matrix extracted from the network analyser. It is expressed in decibel units by the following relationship:

$$S_{11}(dB) = 20\log\left(\frac{onde\_réfléchie}{onde\_incidente}\right) \qquad (2.1)$$

The lower this coefficient, the better the antenna is adapted.

It can be seen that for a reflection coefficient defini a -10 dB, almost 68.4 % of the incident wave is transmitted by the antenna, and for an S.= -40 dB, 99 % of the incident wave is transmitted by the antenna.

The reflection coefficient is also related to the input impedance z_e and the characteristic impedance Z0 of the supply line:

$$\Gamma = S_{11} = \frac{Z_e - Z_0}{Z_e + Z_0}. \qquad (2.2)$$

For: $Z_e = Z_0 \Leftrightarrow S_{11} = 0$, in this case, there is no reflected wave, we are talking about adapting the input impedance to the line.

4.1.2. Bandwidth

Bandwidth (BW) is the range of frequencies determined in дёнёгаl at a standing wave rate *equal to* 2 (equivalent to almost -10 dB of the reflection coefficient curve).

"The range of frequencies within which the performance of the antenna, with respect to certain characteristics, conforms to a specified standard".

The VSWR of an antenna is the main limiting factor for bandwidth.

In our work, we will use the bandwidth where s₁₁ is less than -10 dB (Figure II.2). The bandwidth ratio in % is given by the expression:

$$BW\ (\%) = \frac{la\ bande\ passante}{fréquence\ de\ résonance} \times 100 = \frac{f2-f1}{f0} \times 100 \qquad (2.3)$$

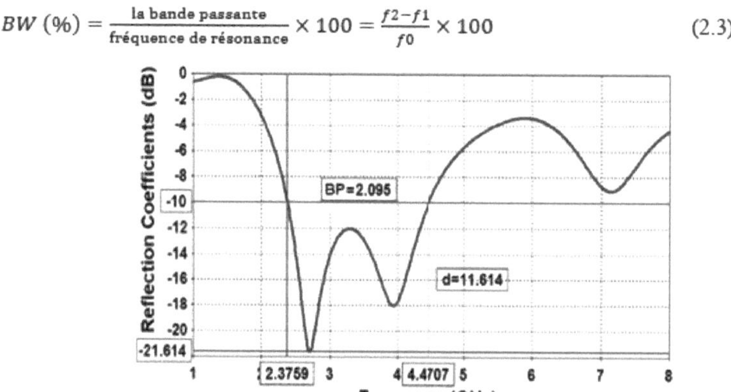

Figure II.2: Bandwidth (s₁₁ less than -10 dB). [6]

The calculation of the bandwidth ratio in figure II.2 is given as a percentage by:

$$BW(\%) = \frac{4.4707 - 2.3759}{f_0} \times 100 \approx \frac{209}{f_0}$$

4.1.3. Voltage Standing Wave Ratio (VSWR)

The ratio of maximum to minimum standing wave diagram values along a transmission line to which a load is connected. The VSWR value varies from 1 (matched load) to infinity for a shorted or open load. For most antennas, the maximum acceptable VSWR value is 2. VSWR is related to the reflection coefficient Γ by:

$$VSWR = \frac{1+|\Gamma|}{1-|\Gamma|} \qquad (2.4)$$

4.1.4. **Input impedance**

"The impedance presented by an antenna at its terminals".

The input impĕdance is a complex frequency function with real and imaginary parts. It is given by:

$$Z_e = Z_0 \frac{1+S_{11}}{1-S_{11}}. \qquad (2.5)$$

Zo: impĕdance characteristic of the supply line.

511: the reflection coefficient.

The input impedance can be presented graphically using the Smith chart.

4.1.5. **Quality factor**

It represents the losses linked to the antenna. A large factor leads to a narrow bandwidth and low efficiency, and is given by the following formula:

$$\frac{1}{Q_T} = \frac{1}{Q_{rad}} + \frac{1}{Q_C} + \frac{1}{Q_d} + \frac{1}{Q_{SW}} \qquad (2.6)$$

Or ; Q_T: Total loss factor.

Q_{rad}: Radiation loss factor.

Q_C: Ohmic loss factor.

Q_d: Dielectric loss factor.

Q_{SW}: Surface wave loss factor.

4.1.6. Performance

Efficiency is the ratio between the energy radiated by an antenna and that supplied by the feed, and is expressed as a function of loss factors. It is given by:

$$\eta = \frac{\dfrac{1}{Q_{rad}}}{\dfrac{1}{Q_T}} = \frac{Q_T}{Q_{rad}} \tag{2.7}$$

4.2. Electromagnetic characteristics [2,7]

4.2.1. Radiation diagram and radiation lobes

This is a graphical representation (in 3-D or 2-D) of the antenna radiation as a function of the angular direction (figure II.3).

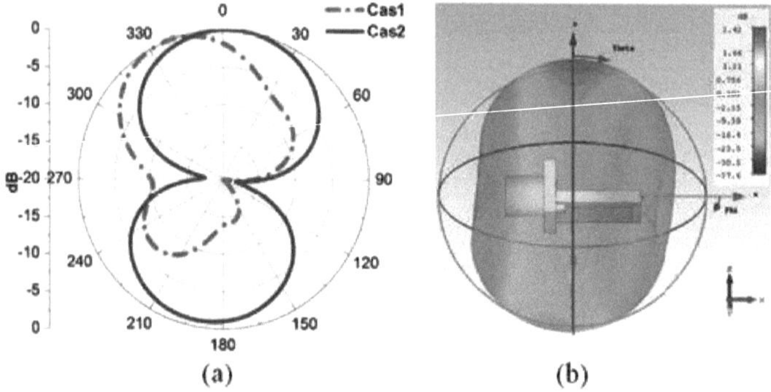

Figure II.3: Radiation diagrams in: a) 2-D and b) 3-D. [8]

The radiation performance of the antenna is gënërally measured and recorded in two main orthogonal planes (such as E-plane and H-plane or vertical and horizontal planes). The diagram is gënërally plotted in polar coordinates. The diagram of most antennae contains a main (major) lobe and several secondary (minor) lobes, called lateral lobes. A lateral lobe appearing in space in the opposite direction to the main lobe is called the arriëre lobe. Figures II.4 and II.5 show a polar and linear diagram, respectively.

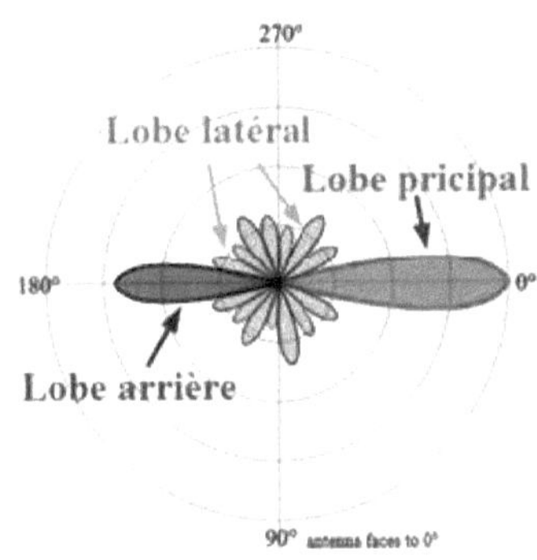

Figure II.4: Main, side and rear lobes.

A radiation lobe *is "a part of the radiation pattern bounded by regions of relatively low radiation intensity".*

A main lobe is defined as *"a lobe of radiation containing the direction of maximum radiation".*

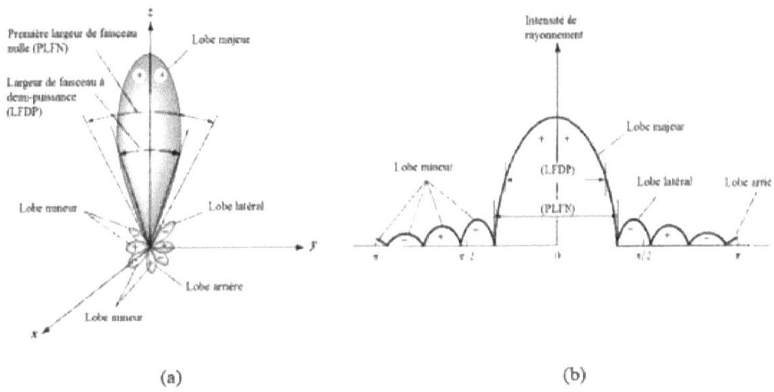

(a)   (b)

Figure II.5: (a) Radiation lobes and beamwidths of an antenna pattern, (b) Linear plot of the power pattern and its associated lobes and beamwidths. [2]

4.2.2. Beam width

Half-Power Beamwidth (HPDW) is defined by the IEEE as follows: *"In a radiation pattern containing the direction of the beam maximum, the angle between the two directions in which the radiation intensity is half the beam value".*

The half-power beamwidth is also referred to as the -3 dB beamwidth (this is illustrated in Figure II.5).

Another important beam width is the angular separation between the first zeros in the diagram, and is known as the First-Null Beamwidth (FNBW).

The HPBW and FNBW widths are both shown in the diagram in Figure II.6. Other beamwidths are those where the diagram is -10 dB from the maximum, or any other value. However, in practice, the term beamwidth, without further identification, generally refers to HPBW.

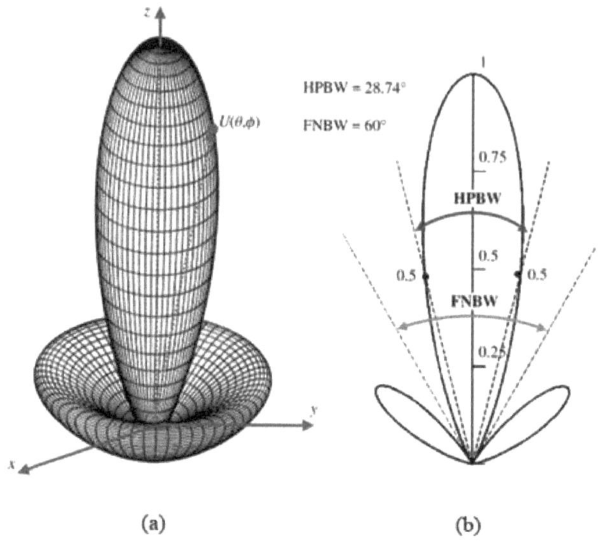

Figure II.6: Power diagrams (linear scale)
of U(9)=cos^2 (9).cos^2 (39) in : (a) 2-D and (b) 3-D. [2]

4.2.3. **Isotropic, directional and omnidirectional diagrams**

The isotropic radiator is defined as *"a hypothetical lossless antenna with equal radiation intensity in all directions"*.

Although it is ideal and not physically possible, it is often used as a reference to express the directional properties of real antennas.

A directional antenna *is "an antenna with the property of radiating or receiving electromagnetic waves more effectively in certain directions than in others"*.

An omnidirectional antenna *is "an antenna having an essentially non-directional pattern in a given plane of the antenna and a directional pattern in any orthogonal plane"*.

The omnidirectional plane is the iorizontal plane, see figure II.7.

Radiation diagram

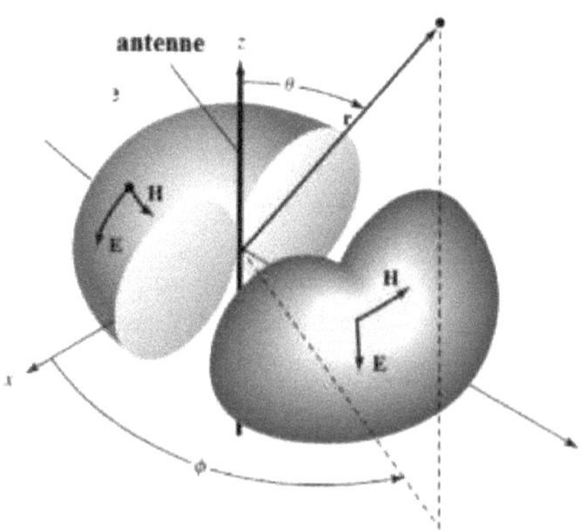

Figure II.7: Omnidirectional antenna pattern. [2]

4.2.4. Main radiation diagram plans

For a linearly polarised antenna, performance is often described in terms of its principal E and H planes (Figure II.8).

The E-plane is defined as *"the plane containing the electric field vector and the direction of maximum radiation"*.

The E plane generally coincides with the vertical plane (XZ plane (elevation plane; ϕ = 0°)).

The H plane is defined as *"the plane containing the magnetic field vector and the direction of maximum radiation"*.

The H plane coincides дёпёга1етеП: with the horizontal plane (XY plane (azimuthal plane; 9 = 90°)).

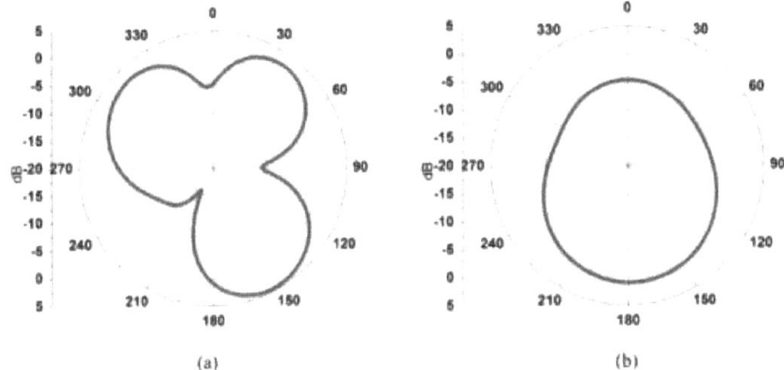

(a)   (b)

Figure II.8: Radiation diagrams; (a) XZ plane, (b) YZ plane. [9]

4.2.5. Field regions

The space surrounding an antenna is generally subdividedë into three regions -as shown in Figure II.9- designated to identify the field structure in each:

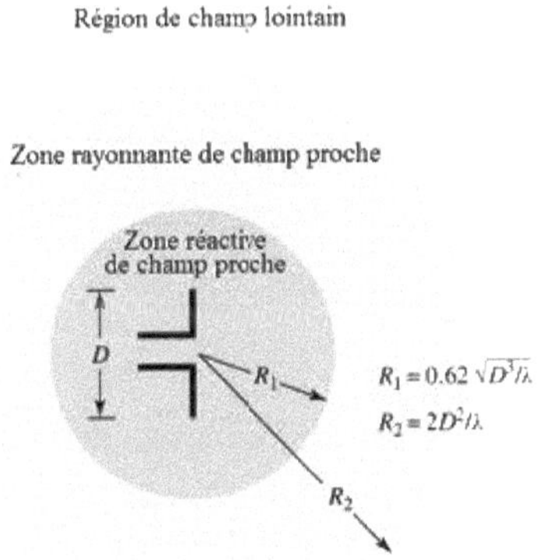

Figure II.9: Field zones of an antenna. [2]

a. Near-field reactive region (Rayleigh zone)

It is defined as *"the part of the near-field region immediately surrounding the antenna in which the reactive field predominates"*.

The outer limit of this region is between $R < R_I$ from the surface of the antenna, with:

$$R_1 = 0.62\sqrt{\frac{D^3}{\lambda}} \qquad (2.8)$$

Or ; X: the wavelength.

D: the largest dimension of the antenna.

b. Near-field radiation zone (Fresnel zone)

It is defined as *"the region of the field of an antenna between the reactive near-field region and the far-field region where the radiation fields predominate and where the angular field distribution depends on the distance from the antenna"*.

The two boundaries (inner and outer) of this region are given by: $R_1 \leq R < R_2$, with:

$$R_2 = 2\frac{D^2}{\lambda} \qquad (2.9)$$

In this region, the field pattern is a function of radial distance and the radial field component can be appreciable.

c. Far-field region (Fraunhofer zone)

It is defined as *"the field region of an antenna whose angular field distribution is essentially independent of the distance from the antenna"*.

The radiation pattern is measured in the far field.

This region is supposëe to exist at distances $R_2 < R < \partial a$

4.2.6. **The amplitude diagram of an antenna**

When the distance of observation varies from the reactive near field to the far field, the amplitude diagram changes shape as a result of the variations in the fields, in both amplitude and phase.

Table II.1: Typical changes in the shape of the antenna amplitude plane since the chamreactive near field to far field.

	Region of reactive fields	Region of radiated fields	Region of distant fields
Field distribution			

From Table II.1, it is evident that in the reactive near-field region, the pattern is more ëtalë and almost uniform, with lëgëres variations.

As the observation is dëplacedëe towards the near-field radiating region, the diagram begins to smooth and form lobes.

In the far-field region, the pattern is well f<m, constitute gënëralement of a few minor lobes and one or more major lobes.

4.2.7. **Radiation power density**

Electromagnetic waves are used to carry information across a wireless medium or guiding structure from one point to another. It is then natural to assume that power and energy are associated with ëlectromagnëtic fields.

The quantity used to describe the power associated with an electromagnetic wave is the instantaneous Poynting vector defined as:

$$W = E \times H \tag{2.10}$$

Or ; W: instantaneous Poynting vector (W/m^2)

E: instantaneous electric field strength (V/m)

H: instantaneous magnetic field strength (A/m)

Since the Poynting vector is a power density, the total power passing through a closed surface can be obtained by integrating the normal component of the Poynting vector over the entire surface. In equation form:

$$P = \iint_S W \cdot dS = \iint_S W \cdot \hat{n}\, da \tag{2.11}$$

Or ; P: total instantaneous power (W)

$\hat{n}$: unit vector normal to the surface

da: infinitesimal area of the closed surface (m^2)

The average Poynting vector (average power density) can be written:

$$W_{av}(x,y,z) = [W(x,y,z;t)]_{av} = \frac{1}{2}\mathrm{Re}(E \times H^{*}) \qquad (\text{W/m}^2) \tag{2.12}$$

If the real part of (E 11*)/2 represents the average power density of an antenna in its far-field region, the imaginary part represents the reactive (stored) power density associated with electromagnetic fields.

The factor 1/2 appears in equation (2.12) because the E and H fields represent crete values. Based on the definition of equation (2.11), the average power radiated by an antenna (radiated power) can be written as ;

$$P_{rad} = P_{av} = \iint_S W_{rad} \cdot dS = \iint_S W_{av} \cdot \hat{n}\, da = \frac{1}{2}\iint_S \mathrm{Re}(E \times H^{*}) \cdot dS \tag{2.13}$$

4.2.8. Radiation intensity

The intensity of radiation in a given direction is defined as *"the radiated power of an antenna per unit of solid angle"*.

Radiation intensity is a far-field parameter, which can be obtained by multiplying the radiation density by the square of the distance.

It is expressed in the following form:

$$U = r^2 W_{rad} \tag{2.14}$$

Or ; U: radiation intensity (W/solid angle unit).

$_{Wrad}$: radiation density (W/m^2).

The total power is obtained by integrating the radiation intensity, given by (2.14), over the entire solid angle of 4π. Done:

$$P_{rad} = \iint_\Omega U\, d\Omega = \int_0^{2\pi}\int_0^{\pi} U \sin\theta\, d\theta\, d\phi \tag{2.15}$$

Or ; $d\Omega = \sin\theta.\, d\theta.$. dtp: solid angle element.

4.2.9. Antenna direction

The directivity of an antenna is defined as *"the ratio of the radiation intensity in a given direction of the antenna to the average radiation intensity in all directions"*. It can be written in the following form:

$$D = \frac{U}{U_0}$$ (2.16)

The average radiation intensity is égale a the total power rayonnée by the antenna dividedée by 4 л. It is given by:

$$U_0 = \frac{P_{rad}}{4\pi}$$ (2.17)

If the direction is not specified, it implies the direction of maximum radiation intensity, which is expressed as:

$$D = D_0 = \frac{U_{max}}{U_0} = \frac{4\pi U_{max}}{P_{rad}}$$ (2.18)

Or ; D: directivity (dimensionless)
D0: maximum directivity (dimensionless)
U: radiation intensity (W/solid angle unit)
Umax: maximum radiation intensity (W/solid angle unit)
U0: radiation intensity of the isotropic source (W/solid angle unit)
Prad: total radiated power (W)
The directivity of an isotropic source is égale a 1, since: $U = U_{max} = U_o$.

4.2.10. Gain

The gain of an antenna (in a given direction) is defined as *"the ratio of the intensity, in a given direction, to the radiation intensity (obtained if the power accepted by the antenna were radiated isotropically). The radiation intensity corresponding to the isotropically radiated power is equal to the power accepted (input) by the antenna divided by 4 л"*.

In most cases, we deal with relative gain, defined as *"the ratio of the power gain in a given direction to the power gain of a reference antenna in its reference direction"*.

The référence antenna is a lossless isotropic source.

$$G = 4\pi \frac{U(\theta, \varphi)}{P_e\left(source\ isotrope\ sans\ perte\right)}$$ (2.19)

P_e *(lossless isotropic source)*

When the direction is not specified, the power gain is gënërally taken in the direction of maximum radiation.

4.2.11. Radiation resistance [4,5]

We define the radiation resistance at a point Q as Q:

$$R_Q = \frac{2P_r}{I_Q^2}$$ (2.20)

Pr: The active power radiated by an antenna.
IQ: The current at a point on this antenna.

4.2.12. Polarisation [10]

An antenna radiates or receives electromagnetic waves. There are three fundamental fagons by which the ëlectromagnëtic wave is radiated, i.e. li^airement (vertically or horizontally), circularly and elliptically [11], see figure II.1o.

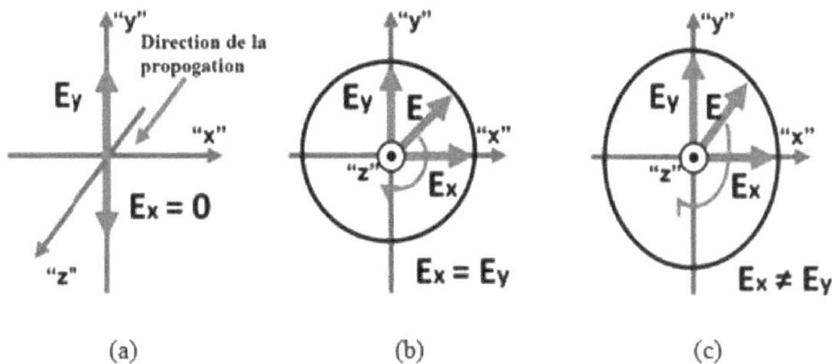

Figure II.10: Polarisations of an antenna: a) Linear, b) Circular, c) Elliptical. [11]

The polarisation of the antenna is defined in consëquence with the polarisation of the electromagnetic wave.

5. Microstrip antennas (MSA)

5.1. Description [12,13]

G.A. Deschamps was the first to introduce the concept of microstrip radiators in 1953, and the first patented documentation of microstrip antennas was produced by Gutton and Baissinot of France in 1955, but the real development didn't take place until the 1970s.

In its most basic form, a microstrip antenna consists of a radiating element (patch) on one side of a dielectric substrate which has a ground plane on the other side as shown in Figure II.11.

The patch is usually made of conductive material such as copper or gold and can take any shape. The radiating patch and power supply lines are usually photo-etched onto the dielectric substrate.

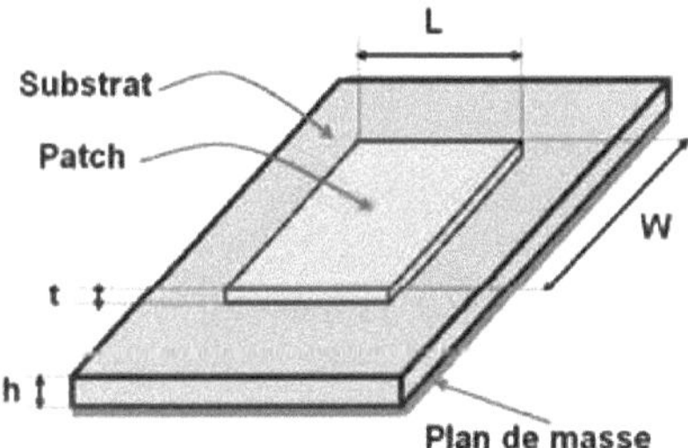

Figure II.11: Structure of the microstrip antenna.

Patch antennas radiate mainly because of the fringe fields between the edge of the patch and the ground plane.

For good 1 antenna performance, a ë thick diëlectrical substrate with a low diëlectrical constant is desirable, as it provides better efficiency, bandwidth and radiation [2].

5.2. Different patch shapes for microstrip antennas

The radiating element (patch) of a microstrip antenna can take several simple or complex forms; among the simple and principal forms are the square or rectangular form, the circular or elliptical form and the triangular form. Of all the existing versions, the rectangular shape is the most widely used, since it offers two degrees of freedom for modifying the antenna's performance [8]. Figure II.12 shows some simple shapes of the radiating element:

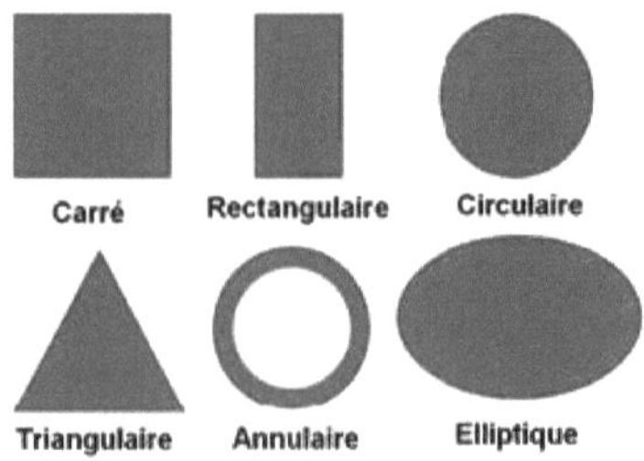

Figure II.12: Different patch shapes.

Variation in the surface area of the radiating element is one of the techniques for improving bandwidth. This variation can lead to a change in the current path and electromagnetic field lines, which in turn changes the operating frequency and radiation pattern of the antenna. There are other shapes (figure II.13) whose patch can be partially modified and sliced by letters (L, H, E, S, V... etc) or can also take on other shapes such as the spiral or meander shape which are involved in the miniaturisation of antennae (see chapter III).

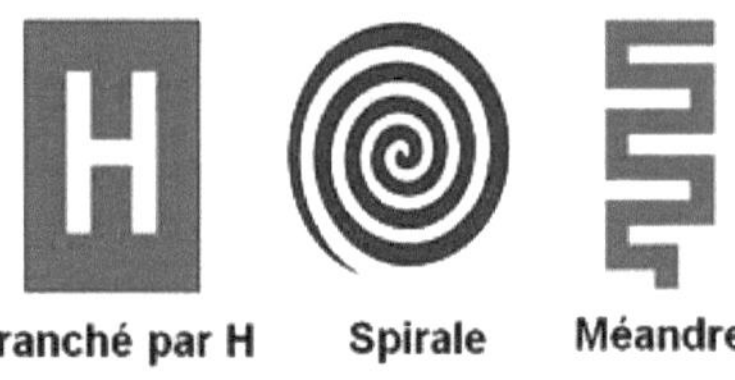

Figure II.13: Other forms of patch.

5.3. Microstrip antenna feed technique [2]

In this section, we will discuss the five techniques widely used to feed microstrip antennas. All techniques are rëalisëes easily using the technology of making imprimës circuits by the photogravure mëthode and cause a bandwidth ёйюHc.

5.3.1. Microstrip transmission line

The microstrip line is a transmission band of very small width gravëed on the same plane with the patch.

The microstrip feed line (Figure II. 14) is easily designed using photo-etched printed circuit technology. The centre conductor of the SMA (for Sub Miniature version A) connection port is soldered directly to this feed line at the antenna end, while the outer conductor is connected to the ground plane. The modëlisation of the antenna is easily rëaliséëe via this type of feed but as a result it presents a narrow bandwidth.

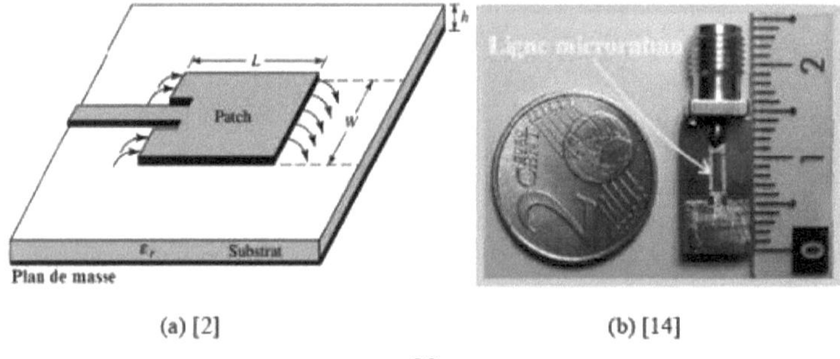

(a) [2]   (b) [14]

Figure II.14: (a) Microstrip feed, (b) Photo of a completed antenna.

The microstrip line feeding technique is favoured for certain antenna array applications (Figure II. 15), but in this case the coupling losses by surface waves increase, which generates interference that disturbs the antenna radiation.

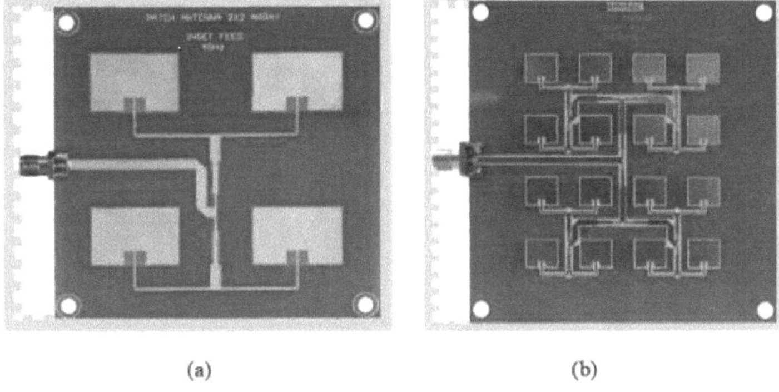

(a)   (b)

Figure II.15: Microstrip line feeding an antenna array of (a) 4 patches,

5.3.2. Coaxial probe

In this type, the substrate must be percë in order to place the central connector of the SMA port (Figure II.16) in the hole and solder it directly to the patch which minimises radiation losses from the line.

Modëlisation of the input impëdance is a little difficult with this technique of feeding by varying the position of the probe at the edges of the radiating l^tement (patch).

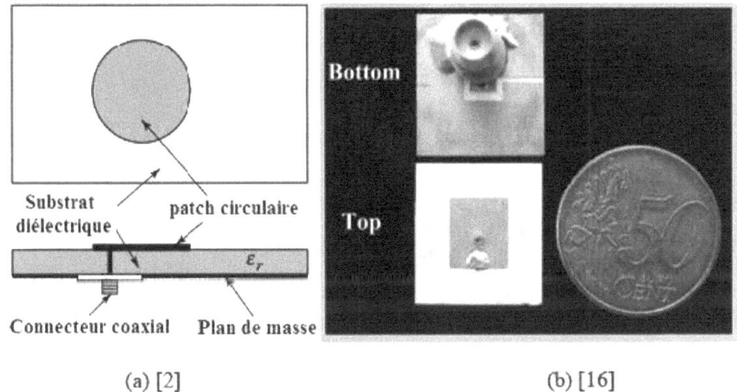

(a) [2]   (b) [16]

Figure II.16: Feed from a coaxial probe, (b) Photo of a completed antenna.

5.3.3.   Aperture coupling (slot)

Two substrates with different permittivities si and £2 are superimposed one on top of the other (sandwich), between which is printed a complete ground plane sliced by a small rectangular aperture (or other shape).

The supply line etched into the underside of substrate 2 emits energy through the slot to substrate patch 1 (Figure II.17).

To modëlise the design, the position of the slit can be centred or dëcalëed below the patch. Only ëlectromagnëtic wave beams directed towards the slit that successfully penetrate substrate 1, other waves are reflected by the presence of mëtalLIque ground plane that separates the two substrates.

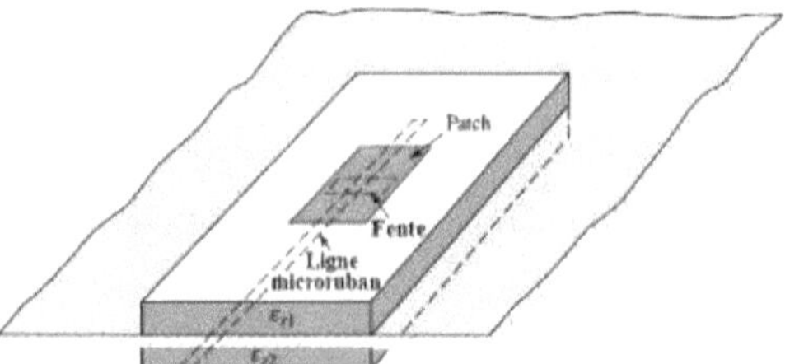

Figure II.17: Aperture-coupled power supply. [2]

5.3.4. Proximity coupling

In this structure, the microstrip supply line (strip) is located between two dielectric layers (Figure II.18):

- a top layer with the patch at the top (antenna substrate);
- a lower layer whose ground plane is at the bottom (feed substrate).

Electromagnetic coupling takes place independently and without contact between the power supply line and the patch, which minimises parasitic radiation.

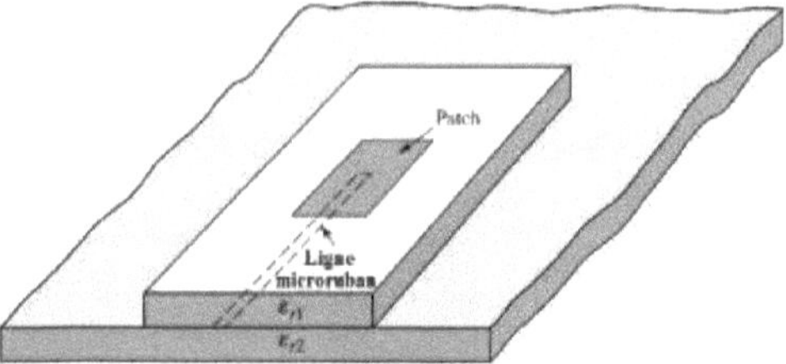

Figure II.18: Proximity-coupled power supply. [2]

Among the four food
s dëcrites at the top:

- Proximity coupling has the widest bandwidth (up to 13%).
- Aperture coupling is the most difficult to model and manufacture.

5.3.5. Coplanar waveguide [17]

In this type of power supply, the coplanar line and the ground plane are aligned on the same side (Figure II. 19), making it possible to place active SMD (surface-mounted component) type components and create reconfigurable antenna or array models.

An antenna fed by a Coplanar Wave Guide (CPW) is simpler than one fed by slot coupling.

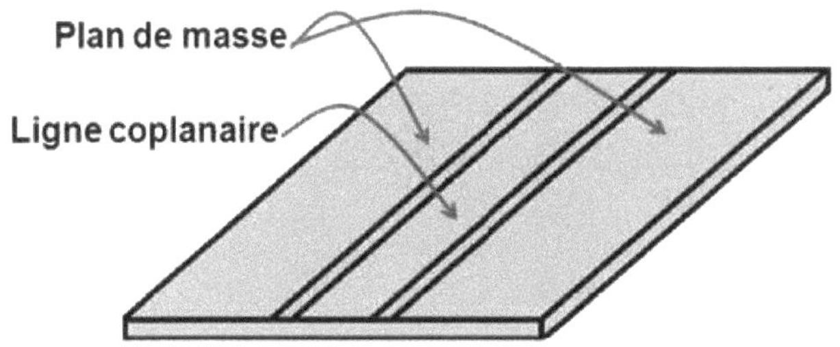

Figure II.19: Coplanar waveguide line.

The advantages of this structure are improved efficiency, wider bandwidth and better isolation between the power supply circuit and the radiating element.

The three models shown in Figure 11.20 present equivalent circuits for each of these power supplies.

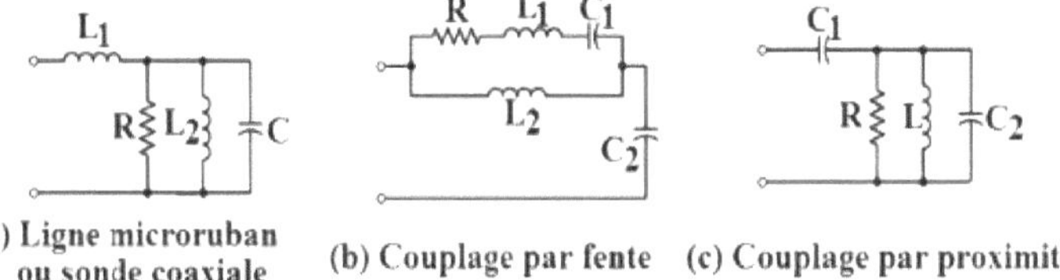

(a) Microstrip line or coaxial probe (b) Slot coupling (c) Proximity coupling

Figure II.20: Equivalent circuits for different configurations.

5.4. Microstrip antenna excitation mode [2]

To determine the dominant mode with the lowest resonance, we need to look at the resonance frequencies. The mode with the lowest resonance frequency is called the dominant mode. Placing the resonance frequencies in ascending order determines the order of the operating modes. For all microstrip antennas h<< L and h<<W.

If: L> W> h, the lowest frequency mode (dominant mode) is TM_{010} whose resonance frequency is given by:

$$(f_r)_{010} = \frac{1}{2L\sqrt{\mu\varepsilon}} = \frac{\upsilon_0}{2L\sqrt{\varepsilon_r}} \qquad (2.21)$$

Or ; u0: the speed of light in free space.

In addition, if: L> W> L/2> h, the next higher (second) order mode is the $TM_0{}^x{}_{01}$ whose resonance frequency is given by:

$$(f_r)_{001} = \frac{1}{2W\sqrt{\mu\varepsilon}} = \frac{\upsilon_0}{2W\sqrt{\varepsilon_r}} \qquad (2.22)$$

However, if: L> L/2>W> h, the second-order mode $is TM_0{}^x{}_{20}$, instead of $TM_0{}^x{}_{01}$, whose resonance frequency is given by:

$$(f_r)_{020} = \frac{1}{L\sqrt{\mu\varepsilon}} = \frac{\upsilon_0}{L\sqrt{\varepsilon_r}} \qquad (2.23)$$

If: W> L> h, the dominant mode is the $TM_0{}^x{}_{01}$ whose resonance frequency is given by

(2.23), whereas if: W> W/2> L> h, the second order is the TM whose resonance frequency is given by (2.23): W> W/2> L> h, the second order is the TM^x_{002}.

The distribution of the tangential electric field along the side walls of the cavity for the modes: TM_{010} , TM_{001} , TM_{020} and TM_{002} is shown in Figure II.21.

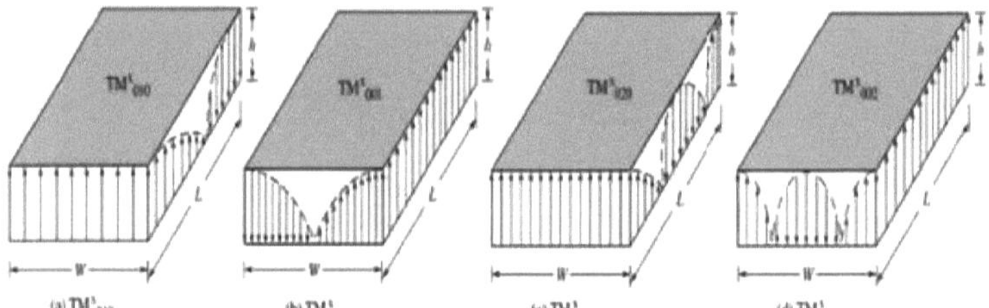

Figure II.21: Field configurations (modes) for a rectangular microstrip antenna.

In all the prëcëdente discussion, he ël.ë assumedë that there is no skin effect (fields along the edges of the cavity). This is not totally valid, but it is a good livpotliese for calculating fields.

5.5. Specific characteristics of microstrip antennas

In this section, we will briefly explain some of the characteristics of microstrip antennas:

5.5.1. Bandwidth [18]

Patch antennas have a narrow bandwidth (for a reflection coefficient <-10 dB) with a bandwidth ratio between 1% and 5% (Impedance BW ratio <5%).

5.5.2. Quality factor [18]

In general, the quality factor is inversely proportional to the size of the antenna in terms of wavelength.

Since the bandwidth is that of a very small microstrip antenna, its Q factor is low. The typical cpЫПë factor of a patch is between 50 and 75.

5.5.3. Radiation pattern and directivity [19]

The tvpic radiation diagrams in the E and H planes are shown in Figure II.22. If the ground plane is finite, leakage will occur towards the lower half-plane.

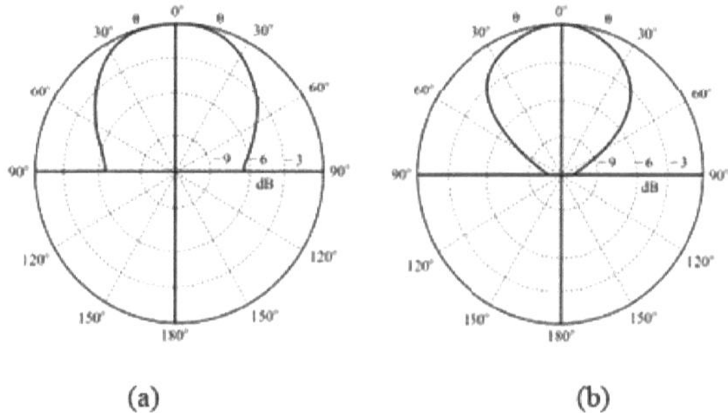

(a)   (b)

Figure II.22: TVp radiation patterns of a rectangular patch antenna for (a) E-plane and (b) H-plane.

antenna for (a) E plane and (b) H plane [19].

5.5.4. Gain [20]

A microstrip antenna built with a single radiating element (patch) will have a maximum gain of around 6 dBi.

5.5.5. Efficiency

Microstrip antennas are characterised by low efficiency.

5.5.6. Polarisation [20]

Most microstrip antennas are designed with linear polarisation. There are many applications, such as the creation of 3D films etc, where it is more reliable to use circular polarisation.

Circularly polarised microstrip antennas can be classified into three types: single feed, dual feed and sequential rotation.

Figure II.23 shows four unique feed designs. Figure II.23 (a) and (b) show, respectively, a nearly square patch and a nearly circular (elliptical) shape. Figure II.23 (c) and (d) show, respectively, a square patch with truncated corners and a circular patch with indentations. Although coaxial probes are shown in this figure, patches can also be fed by a microstrip line or by coupling through an aperture. This design is simple, but suffers from a very narrow bandwidth for thin substrates.

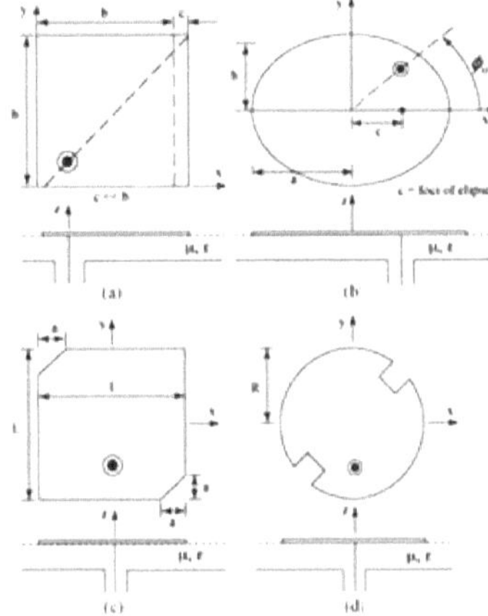

Figure II.23. Circular polarisation of patches: (a) almost square; (b) elliptical.
(c) square with truncated corners. (d) circular with indentations [20].

1.6. **Advantages and disadvantages of microstrip antennas [21-23]**

1.6.1. Benefits

Microstrip antennas offer a number of important advantages, including

- They are easy to manufacture using printed circuit technology and are inexpensive to produce;
- They support both linear and circular polarisation;
- Can be integrated with other hyperfrequency modules;
- Radiant structure, no cavity required;
- Low weight, minimum footprint.

1.6.2. Disadvantages

This technology suffers from the following limitations:

-       Narrow bandwidth ;
-       Lower power gain (-6 dB) ;
-       Surface wave and ohmic loss ;
-       Stray radiation and orientc in a half-plane ;
-       A high power supply will cause the antenna and its connections to burst.

6.   Delectric Resonator **Antenna** (DRA)

The aim of this section is to provide an understanding of the operation of rectangular RD antennas (RDRA), their fundamental and secondary modes, the expression of their resonance frequencies, the field shapes inside the resonator and the radiation patterns corresponding to each mode.

6.1. Description [24,25]

The electrical resonators using high-permittivity materials were developed for microwave circuits, such as filters or oscillators.

When a dielectric resonator is placed in a complementary open environment (without a shielding cavity), the Q factor of the lowest modes is greatly reduced to around 10 to 100, because microwave power is lost through the effect of radiated fields. This fact was realised by Richtmeyer [26] in 1939 and makes dielectric resonators useful as antenna elements.

In the late 1960s, the development of low-loss cëramic matërials paved the way for their use as high-factor elements. The possibility of constructing small dielectric resonator antennas was first reported by Sager and Tisi [27] in 1968.

In the early 1980s, S. A. Long was the first to use multiple feeding mechanisms to excite different modes of a dielectric resonator antenna (DRA).

An experimental array of dielectric resonators excited by a dielectric guide was reported by Birand and Gelsthorpe [28] in 1981, although the first experimental and theoretical study of a dielectric resonator antenna had to wait until 1983 [29].

The determination of the input impedance, the Q factor and the fields inside the resonator was achieved during the 90s by applying analytical and/or numerical techniques.

Kishk, Junker, Glisson, Luk, Leung, Petosa and so on have described a significant amount of DRA analysis. A review of the design and calculation of the modes and radiation characteristics of dielectric resonator antennas has been published by Mongia and Bhartia [30].

Current literature focuses on miniaturised designs of RD antennas to improve antenna bandwidth and address wireless handheld applications.

6.2. **Different shapes of RD antennas [25,31].**

RDs come in a variety of shapes, including cylindrical (the most popular), ring, tubular, spherical and parallelepipedic.

Figure II.24 shows cylindrical, rectangular, hemispherical, low-profile circular disc, low-profile triangular and spherical cap RDs.

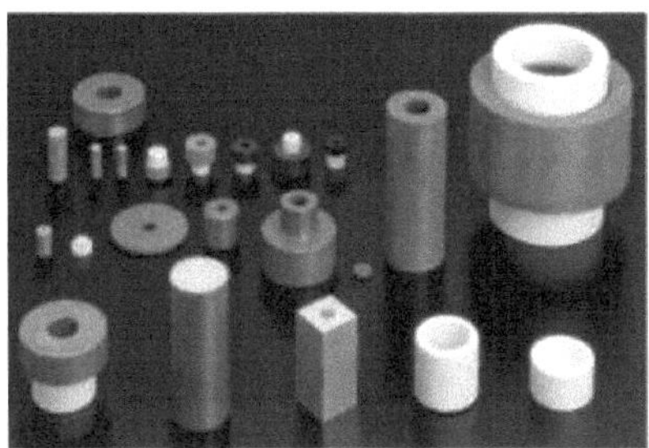

Figure II.24: Dielectric resonators of different shapes. [32]

By modifying the geometry of the RD, it is possible to modify and control the excitation of the antenna and obtain an expected radiation pattern [33]. The internal ëlectric field of the RD can be controlled with variations in shape [34] and conversely by making controlled modifications to the shape, it is possible to adjust the performance of the antenna by modifying the internal electric field.

There is a direct correlation between the shape of the DRA and its performance. Since the first papers were published on the concept of DRA, much work has been done on the basic forms of DRA.

6.3. Type of modes (TE, TM, HEM) [35]

There are four types of electromagnetic waves:

1. Transverse electric and magnetic mode (TEM)
2. Transverse electrical mode (TE)
3. Transverse magnetic mode (TM)
4. Hybrid electric and magnetic mode (HEM) or odd HE and even EH

Mode propagation depends mainly on the following configuration:

5. Excitement
6. Dimensions
7. Coupling
8. Environment
9. Excitation point
10. Input impedance

6.4. Resonance modes [35]

With knowledge of the modes, the radiation characteristics of an antenna can be predicted and the designer can therefore provide a correction in the antenna design.

In the RDRA, resonance modes represent radiation phenomena using E and H field models. The resonance modes are E and H field patterns within the RDRA. Figure II.25 shows that electric fields are always associated with magnetic fields and vice versa.

The modal field equations are developed using Fourier-based functions of cosine or sine terms appearing on the basis of the RDRA boundary conditions, i.e. the six RDRA walls can be PMC, PEC or any combination of these walls.

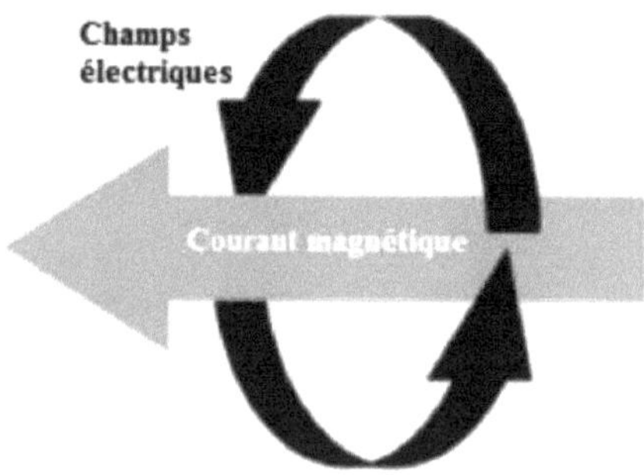

Figure 25:   5: Electric fields and associated magnetic fields

Consequently, the resonant modes provide a physical insight into the radiant phenomena taking place inside the RDRA. The resonant modes form a set of orthogonal functions for calculating the total current over the surface of the RDRA.

Figure II.26 shows the configuration of the resonant modes generated in RDRA. The wave can only propagate if the wave vector k> kc, where kc is the cut-off frequency. The lowest resonance is called the dominant mode.

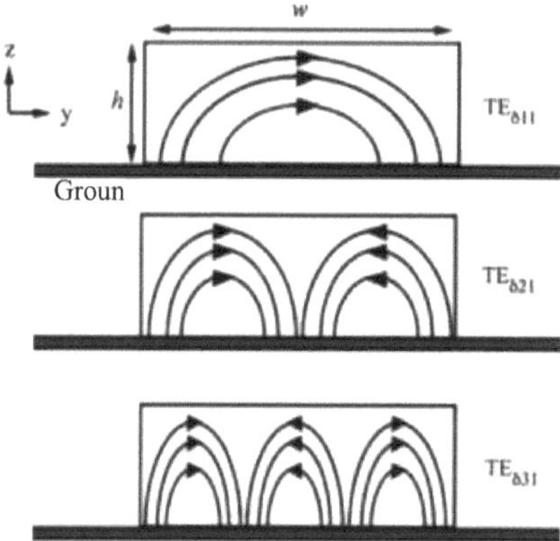

Figure 26:   6: Resonance modes in the yz plane. [35]

6.5. Common feed technique for RD antennas [24].

Multiple feed mechanisms are used to excite different modes of resonators. This sub-section will summarise the most widely used excitations:

6.5.1. Excitation by coaxial probe

The probe can be located within or adjacent to the ARD. Within the RD antenna, good coupling can be achieved by aligning the probe along the electric field of the ARD mode as shown in Figure II.27.

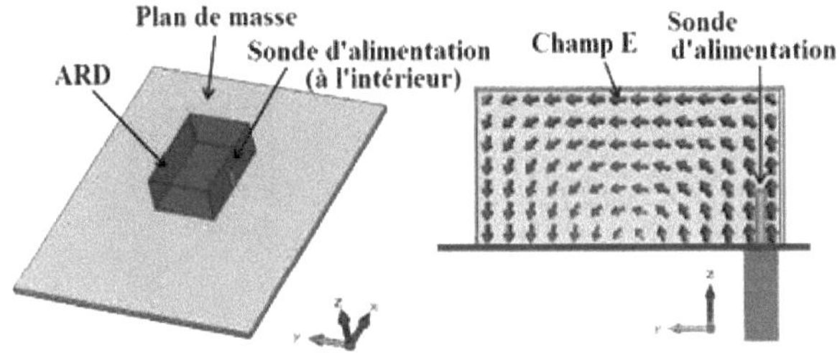

Figure II.27: Coaxial probe coupling the E field. [24]

The adjacent position is used to couple the magnetic field of the DRA mode (Figure II.28). In both cases, the probe excites the TE$_{111}$ fundamental mode of the rectangular DRA.

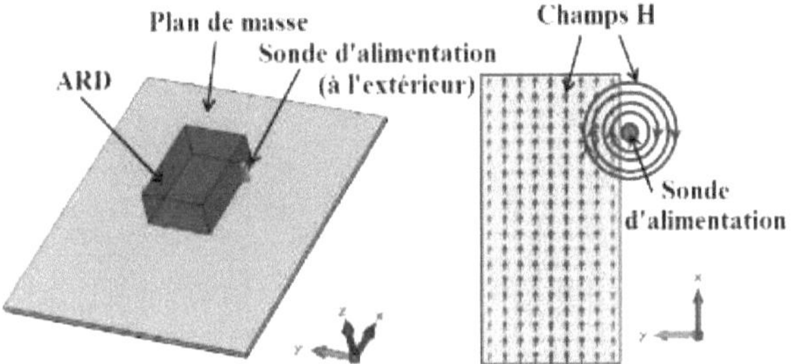

Figure II.28: Coaxial probe coupling the H field. [24]

When the excitation probe is inside the resonator, the air gap (between the excitation probe and the didlectrical material) leads to a lower effective didlectrical constant a $_r$, which results in both a decrease in the Q factor and a shift in the resonance frequency [36,37].

The location of the probe allows the intended excitation mode to be selected and the mode coupling can be optimised by adjusting the length and height of the probe.

6.5.2. Microstrip feed line and coplanar waveguide

The principle is similar with coaxial probe excitation. A microstrip line placée near the antenna a RD generates the following phënomënes:

- can couple the magnetic field of the mode of the latterière ;
- can affect the polarisation of the antenna;
- increase stray radiation.

This could be reduced by placing the line under the resonator as shown in Figure II.29.a.

Another manière is to replace the microstrip line with a coplanar waveguide, Figure II.29.b shows a rectangular ARD excited by a coplanar waveguide.

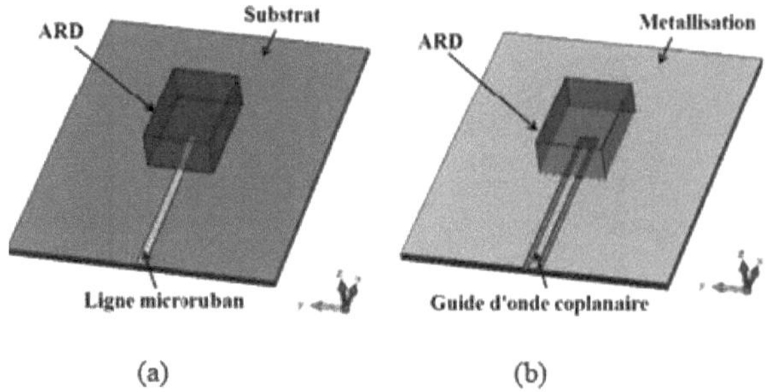

Figure II.29: Feed line; (a) microstrip and ; (b) coplanar waveguide. **[24]**

In both cases, the coupling mode can be optimised by changing the position of the resonator and/or its dielectric permittivity.

These excitation methods perturb the ARD modes by introducing electrical boundary conditions. This phenomenon is all the more sensitive when the antenna is miniature.

6.5.3. Coupling by opening

The method of exciting an RD antenna by aperture coupling consists of acting through an aperture in the ground plane.

Figure II.30 shows an example of the TE_{111} mode excitation of a rectangular ARD with a rectangular aperture. To obtain appropriate coupling, the aperture must be placed in a strong magnetic region of the ARD. Feeding the aperture with a microstrip line is a current approach [38,39].

The main dimension of the aperture should be around $x_{g/2}$, which is very problematic at low frequencies.

In addition to these multiple feeding methods, the choice of different forms of DRA represents another degree of flexibility and versatility.

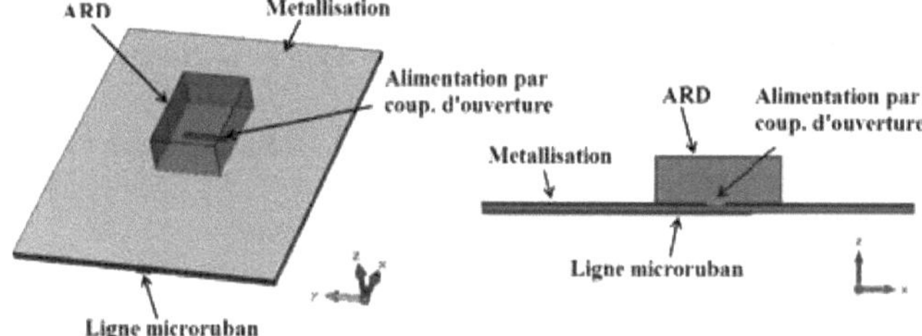

Figure II.30: Aperture coupling the TE_{111} mode of rectangular ARD. **[24]**

6.6. Characterisation of the resonance modes of rectangular DRA [24].

The rectangular ARD is characterised by its length a, width b and height d. It has three independent lengths, giving it one degree more liberty than the cylindrical ARD.

The modéle used to analyse rectangular ARD is the dielectric waveguide model [40-42].

In Figure II.31, the bottom surface is an electrical wall (since the DRA is mounted on a ground plane). The top surface and two sides of the ARD are assumed to be perfect magnetic walls, while the other two are imperfect magnetic walls (since the case considered is realistic).

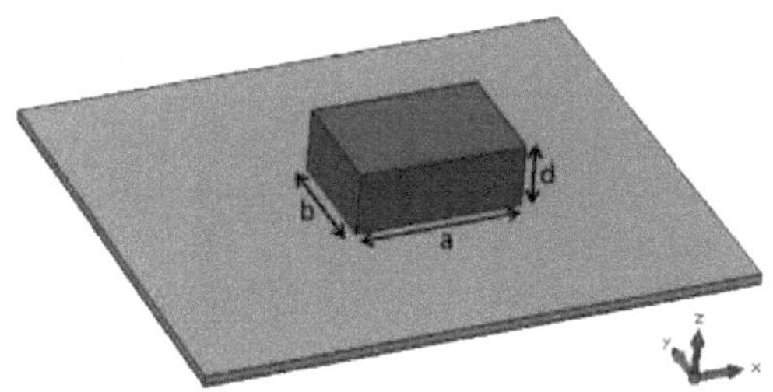

Figure II.31: Rectangular RD antenna. **[24]**

The modes in an isolated rectangular dielectric resonator can be divided into two categories: TE and TM modes, but when the ARD is mounted on a ground plane, only TE modes are typically excited. The fundamental mode is TE₁₁₁. As the three dimensions of the DRA are independent, the TE modes can be located in the following three directions: x, y and z.

Referring to the cartesian coordinate system shown in figure II. 31, if the dimensions of RD are such that: a> b> d, the modes in resonance frequency order are: TE_{111} , TE_{111} and TE_{111} . The analysis of all the modes is similar.

The *TE* mode example₁₁₁ is discussed in [40], the field components inside the resonator and the resonant frequencies are presented analytically. The CST MS software can be used to visualise the E and H fields, as shown in figure 11.32.

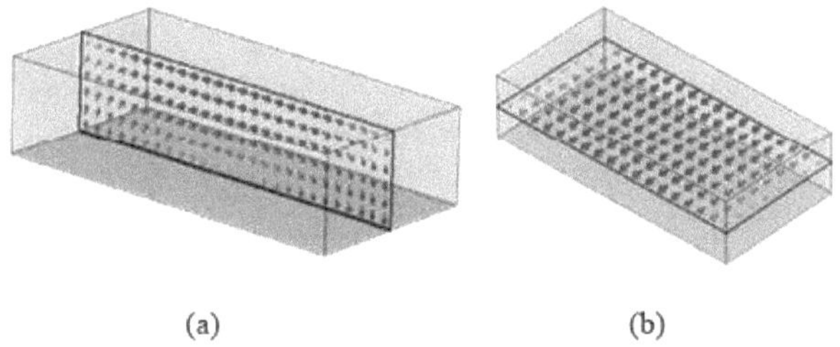

(a)   (b)

Figure II.32: (a) E-field and (b) H-field of the TE₁₁₁ mode **[24].**

The characteristic equations of the ARDR are given below:

$$\begin{cases} k_0^2 = k_x^2 + k_y^2 + k_z^2 \\ k_0 = \dfrac{\omega_0}{\upsilon} = \dfrac{2\pi f_0 \sqrt{\varepsilon_r \mu_r}}{c} \end{cases} \qquad (2.24)$$

Or ; k_0: free space wave ;

k_x, k_y, k_z: propagation constants in the x, y and z directions, respectively.

The definition of resonance frequencies is given below:

$$f_0 = \frac{c}{2\pi \sqrt{\varepsilon_r \mu_r}} \sqrt{k_x^2 + k_y^2 + k_z^2} \qquad (2.25)$$

Ou: $k_x = \dfrac{m\pi}{a}, k_y = \dfrac{n\pi}{b}, k_z = \dfrac{p\pi}{d}$

It is found by solving the following transcendental equation -when the fields propagate in the Z direction-:

$$k_z \tan\left(\frac{k_z a}{2}\right) = \sqrt{(\varepsilon_r - 1)k_0^2 - k_z^2} \tag{2.26}$$

Using this model, the values of the resonance frequencies calculated are close to the values measured if a high sr permittivity is used. A frequency shift appears for low sr but the model remains a good approximation method.

The expression for the resonance frequency of an RDRA can be given as follows: [35]

$$f_{r(m,n,p)} = \frac{c}{2\pi\sqrt{\varepsilon\mu}} \sqrt{\left(\frac{m\pi}{a}\right)^2 + \left(\frac{n\pi}{b}\right)^2 + \left(\frac{p\pi}{d}\right)^2} \tag{2.27}$$

6.7. Electrical and electromagnetic characteristics of DRAs

6.8. 1. Electrical characteristics [43, 44]

a. Resonance frequency

Fundamental electromagnetic field theory suggests that the modes of a dielectric waveguide can be TE, TM or hybrid. Therefore, for the analysis of RDRA modes, only TEmnl modes will be discussed, the indices m, n and l indicating the order of variation along the x, y and z directions of the Cartesian coordinate system, respectively.

The normalized frequency is defined as:

$$F = \frac{2\pi\omega f_0 \sqrt{\varepsilon_r}}{c} \tag{2.28}$$

b. Q factor and bandwidth

The choice of mode to use for antenna applications depends on the Q factor that is used to assess the performance or quality of a resonator.

The Q factor or quality factor is a measure of the operating bandwidth and is defined by:

$$Q = \frac{\omega_0 U}{P} \tag{2.29}$$

Or ; ω_0: resonance pulse;

U is the stored energy;

P is the power dissipation.

When the resonator oscillates freely without being driven by an external source, we refer to this as the unloaded Q. When an external source is permanently connected to the resonator, the appropriate Q is the Q load.

For antennas, the important Q factor is Qrad, or the power dissipated from the P term in equation 2.29 is the radiated power. Van Bladel [45] stated that Qrad is proportional to r, for diëlectric resonators and this relationship -for o,- e l eyë- is given as follows:

$$Q_{rad} \alpha \in_r^p \tag{2.30}$$

Or: P = 1.5 for modes that radiate like a magnetic dipole, and P = 2.5 for modes that radiate like an electric dipole or like a magnetic quadripole.

The bandwidth is defined as the frequency bandwidth in which the input standing wave ratio (VSWR) of the antenna is less than a specified value S. The greater the bandwidth, the greater the coverage over the frequency space that the antenna can be used.

Mongia [46] has presented an analytical expression for the bandwidth, this relationship is given as follows:

$$BW = \frac{S-1}{Q_u \sqrt{S}} \tag{2.31}$$

Or ; Q_u: Q decharge factor.

Dielectric resonator antennas have a negligible dielectric and conductor loss compared to their radiated power. Consequently, the Q radiated,

$$Q_{rad} \cong Q_u \tag{2.32}$$

P values are only valid for very high permittivity values (n $_{r>100}$). Lower order modes generally have a lower Q_{rad}, making them more suitable for practical applications. The Q_{rad} factor is important because a low Q_{rad} would indicate a wide bandwidth but a high Q_{rad} would indicate a low operating bandwidth.

The final expressions for total stored energy W and radiated power P_{rad} are [47]:

$$Q = \frac{2\omega_0 W_e}{P_{rad}} \tag{2.33}$$

Or ; P_{rad} and W_e: radiated power and stored energy, respectively.

These quantities are given by:

$$W_e = \frac{\varepsilon_0 \varepsilon_r abdA^2}{32}\left(1+\frac{\sin(k_z d)}{k_z d}\right)(k_x^2 + k_y^2) \tag{2.34}$$

$$P_{rad} = 10k_0^4 |P_m|^2 \tag{2.35}$$

Where: Pm is the magnetic dipole moment of the ARD:

$$P_m = -\frac{8jw\varepsilon_0(\varepsilon_r - 1)A}{k_x k_y k_z}\sin\left(\frac{k_x d}{2}\right)\hat{x} \tag{2.36}$$

The Q factor погтаНзё (Q_e) is dëfmi as:

$$Q_e = \frac{Q}{\varepsilon_r^{3/2}} \tag{2.37}$$

6.9. Advantages of DRA antennas [35,48,49].

Diëlectric resonators are efficient radiant ëlements, and their main advantages include the following:

- Low conduction losses due to the use of diëlectrical matëriau, therefore a ëlevës efficiency and Q factor ;
- Compact in size and portable;
- The choice of a more ëlevëe permittivity DRA can considerably reduce the size of the $_{Ag\,hy/sT}$ order;
- Easy to make ;
- No frequency drift due to tempërature change;
- High-power handling capability ;
- Gain ëkyë and bandwidth ëlevëe ;
- Can be пПёдге with MMIC ;
- Various modes can be дёиёlё by varying one of the aspect ratios;
- simple coupling scbumas ;

- The bandwidth can be varied by choosing the diëlectrical constant;
- Variëtë of possible shapes;
- possibility of feeding them with all the classic mëthodes.

7. Comparison between MSA/DRA branches [50,51].

In rëfërence [50], the performance of the DRA antenna and the MSA antenna were compared at millimetre wave frequencies by measuring a cylindrical DRA and a circular MSA disc. These two antennas were built on the same substrate and fed by microstrip lines of the same length using quarter-wave (X/4) transformers.

The measurement results show that:

a) The design and manufacture of the DRA is a little more complex than for the MSA.

b) DRA may be more difficult to integrate into planar circuits due to its three-dimensional profile (bonding to the substrate).

c) Improved performance in terms of bandwidth and radiation efficiency is recorded for DRA antennas compared with MSA antennas:

- the bandwidth of the DRA is wider than that of the MSA ;
- while the radiation diagrams are lëgërementally wider than those of the MSA ;
- the diëlectrical resonator antenna can radiate more efficiently than the microstrip antenna (the absence of conductive loss).
- DRA can be considered a promising alternative for antennas applied at millimetre wave frequencies.

The rëfërence [51] gives some practical guidance on the choice or recommendation of an appropriate ëlë radiating ëment, either microstrip or RD, depending on the practical requirement and design spëcification, but this time with different mëfeed mechanisms.

The results obtained with the two branches (MSA and DRA) are summarised in the tables below:

Table II.2: Comparison of bandwidth characteristics (measured and <u>simulated</u> parameters).

Bandwidth (S11<-10 dB)		Cylindrical microstrip antennas			Cylindrical RD antennas		
		Coaxial probe	Microstrip line	Slot	Coaxial probe	Microstrip line	Slot
Absolute value (MHz)	Simulated values	90	80	100	330	415	355
	Measured values	90	90	110	380	350	210
In percent (%)	Simulated values	2.2	2	2.5	8.3	9.7	9.2
	Measured values	2.2	2.2	2.9	9.5	8.3	5.4

Table II.3: Comparison of primary antenna radiation characteristics (<u>values in brackets are simulated prediction parameters</u>).

Radiation parameters	Cylindrical microstrip antennas			Cylindrical RD antennas		
	Coaxial probe	Microstrip line	Slot	Coaxial probe	Microstrip line	Slot
Plan width E	69 (73)	87 (90)	82.5 (78)	78 (110)	69 (110)	110 (99)
beam (deg) Plan H	80 (70)	75 (89)	84 (88)	80 (70)	81 (90)	99 (90)
Gain (dBi)	6.5 (6.6)	6.1 (6.4)	5.8 (5.9)	5.4 (5.4)	5.2 (5.6)	4.8 (5.1)
Efficiency (%)	87	80	82	96	92	93

Table II.4: Cylindrical DRA performance for different dielectric material permittivities.

Type of power supply	Antenna characteristics	permittivity of dielectric material		
		20	10	6
Coaxial probe	Freq. Resonance (GHz)	2.88	3.95	4.95
	Bandwidth (%)	3.48	8.3	12.82
	Gain (dBi)	3.72	5.3	5.64
Slot	Freq. Resonance (GHz)	2.71	3.86	4.93
	Bandwidth (%)	3.21	9.2	12.64
	Gain (dBi)	3.7	5.1	5.64
Microstrip line	Freq. Resonance (GHz)	3.04	4.25	-
	Bandwidth (%)	4.91	9.7	-
	Gain (dBi)	4.08	5.66	-

Commentaries and comparisons between the two technologies were reported in [51], following analysis of the measurements:

- The DRA has a head start over the MSA in terms of wider bandwidth and more élevée efficiency, but at the cost of a compromise with the gain value of around 1 dB. It should be noted that the characteristics of the two types of antenna depend on their respective dïelectrical properties.

- DRA or MSA intëgrës require microstrip or aperture power supplies, which are superior in terms of XP performance. However, if efficiency is the issue, coaxial probe power supplies are preferred.

- Although DRA appears to outperform MPA in some ëgards, the mechanical treatment and stabile nature of MPA are advantages over DRA.

8. Applications of MSA/DRA antennas [52].

The antennas are used in the following areas:

- Radio frequency spectrum management
- Communications systems
- Tëlëvision and FM broadcasting
- Radar
- Tëlëdëtection
- Radio astronomy

9. Conclusion

In this chapter we have explained the different feeding techniques for microstrip and rectangular dïelectrical resonator antennas.

By comparing the two technologies, it is possible to select the one that is most compatible for mobile and wireless communication systems.

In this chapter, the main forms of antennas and their ëtës electrical and ëlectromagnëtic properties have ëlësentë and dëcribed with all the fundamental parameters. This should be sufficient to understand the general operation of planar antennas used in wireless communication systems.

Bibliography of Chapter II

[1] Zhi Ning Chen, Michael Yan Wah Chia, "*Broadband planar antennas-design and applications*", Pp. 5, Wiley&Sons, Inc, 2005.

[2] Constantine A. Balanis. *Antennas Theory - Analysis and Design. 3rd Edition.* John Wiley&Sons, Inc, 2005.

[3] PhD These, Lei Xing, "*Investigations of Water-Based Liquid Antennas for Wireless Communications*", University of Liverpool, September 2015.

[4] These de doctorat, Mondher LABIDI, ' *Conception et application des metamateriaux pour des circuits RF*, Ecole Superieure des Communications de Tunis, 2012.

[5] These de doctorat, М|скаёl Jeangeorges, '*Conception d'antennes miniatures integrees pour solutions RFSiP*', Electronique, Universite de Nice-Sophia Antipolis, 2 Dec. 2010.

[6] F. Chetouah, N. Bouzit, I. Messaoudene, S. Aidel, M. Belazzoug, Y. Braham Chaouche, "*Annular Dielectric Resonator Loaded with Strip Loop Antenna for Tri-band Applications*", Pp. 862, 13th International Wireless Communications and Mobile Computing Conference (IWCMC), , 26-30 June 2017, Valencia, Spain.

[7] Lokman Kuzu, Erdogan Alkan, "*Microwave Planar Antenna Design*", ELE 791 Project Report, Syracuse University, Spring 2002.

[8] Farouk Chetouah, Salih Aidel, Nacerdine Bouzit, Idris Messaoudene, "*A miniaturized printed monopole antenna for 5.2-5.8 GHz WLAN applications*", Int J RF Microw Comput Aided Eng, 2018.

[9] F. Chetouah, N. Bouzit, I. Messaoudene, S. Aidel, M. Belazzoug, Y. Braham Chaouche, "*Miniaturized Wideband Printed Rectangular Patch Antenna for X- and Ku-Bands*", Pp. 847, 13th International Wireless Communications and Mobile Computing Conference (IWCMC), , 26-30 June 2017, Valencia, Spain.

[10] PhD These, Maria De Los Angeles Castillo Solis, '*Dielectric resonator antennas and bandwidth enhancement techniques*', University of Manchester, 2014.

[11] Kraus,J.D., [1950], "*Antennas*," New Yor-Toronto-London Mc Graw-Hill Book company, Electrical and Electronic Engineering Series, Federick Emmos Terman, Consulting Editior; W.W Harman and J.G Truxal, Associate Consulting Editors; ISBN 07-035410-3; pp 465.

[12] Rachmansyah, Antonius Irianto, A. Benny Mutiara, "*Designing and Manufacturing Microstrip Antenna for Wireless Communication at 2.4 GHz*", International Journal of Computer and Electrical Engineering, Vol. 3, No. 5, Oct. 2011.

[13] Ribhu Abhusana Panda, Upasana Patnaik, Nibedita Bisoyi, Kiran Tripathy, "*Microstrip Patch Antenna Design at 5.2GHz*", IJESC, Vol. 7, Issue No.4, 2017.

[14] F. Chetouah, N. Bouzit, I. Messaoudene, S. Aidel, M. Belazzoug, Y. B. Chaouche, "*Miniaturized printed rectangular monopole antenna with a new DGS for WLAN applications*", International Symposium on Networks, Computers and Communications (ISNCC), Marrakech, Morocco, 16-18 May 2017.

[15] https://qph.ec.quoracdn.net/main-qimg-329c8b8d3ce6e67396b1f30208e89b00

[16] A. A. Salih, M. S. Sharawi, "*Highly miniaturized dual band patch antenna*", 10th European Conference on Antennas and Propa-gation (EuCAP), Davos, Switzerland, 10-15 April 2016.

[17] http://qucs.sourceforge.net/tech/node86.html

[18] Robert A. Sainati, "*CAD of Microstrip Antennas for Wireless Applications*", Artech House Antennas and Propagation Library, Pp. 5, 1996.

[19] Yi Huang, Kevin Boyle, "*Antennas From Theory to Practice, Chapter 5: Popular Antennas*", Pp. 187, John Wiley and Sons, Ltd, 2008.

[20] Kai-Fong Lee, Kin-Fai Tong, "*Microstrip Patch Antennas-Basic Characteristics and Some Recent Advances,*" Proceedings of the IEEE, Vol. 100, No. 7, Pp. 2169 - 2180, July 2012.

[21] Zhi Ning, Chenand Michael, Y. W.Chia, "*Broadband Planar Antennas: Design and Applications, Chapter two: Broadband Microstrip Patch Antennas*", John Wiley & Sons, 2006.

[22]   https://prezi.com/afmbgt_x7lfl/les-antennes-patch/

[23]   Memoire de maitre es sciences '*Etude et realisation des antennes ultra large bande a double polarisation* ', Rabia Yahya, Universite du Quebec INRS- EMT, 2011.

[24]   Marius Alexandru Silaghi, "*dielectric material*", Section 1: Chapter 2: *Dielectric Materials for Compact Dielectric Resonator Antenna Applications*, Pp. 27-30, L. Huitema, T. Monediere, InTech, Croatia, 2012.

[25]   Hari Singh Nalwa, "*Handbook of Low and High Dielectric Constant Materials and Their Applications*", Stuart Penn, Neil Alford, 'Chapter 10: *ceramic dielectrics for microwave applications*', Academic Press, 1999.

[26]   R. D. Richtmeyer, J. Appl. Phys. 10, 391, 1939.

[27]   O. Sager, F. Tisi, Proc. IEEE, 56, 1593, 1968.

[28]   M. T Birand, R. V. Gelsthorpe, Electron. Lett. 17, 633, 1981.

[29]   S. A. Long, M. McAllister, L. C. Shen, IEEE Trans. Antennas Propagat, 31, 406, 1983.

[30]   R. K. Mongia, P. Bhartia, Int. J. Microwave Millimeter-Wave CAE, 4, 230, 1994.

[31]   These of PhD, Maria De Los Angeles Castillo Solis, "*Dielectric Resonator Antennas And Bandwidth Enhancement Techniques*", University of Manchester, 2014.

[32]   These de Doctorat en Electronique, Hedi Ragad, '*Etude et conception de nouvelles topologies d'antennes a resonateur dielectrique dans les bandes UHF et SHF*', 22 Nov 2013, Tunisia.

[33]   K. S. Ryu, A. A. Kishk, "*Ultra-Wideband Dielectric Resonator Antennas,*" International Workshop on Antenna Technology (iWAT), Pp. 1 - 4, 2010.

[34]   Denidni T. A., Weng Z., and Niroo-Jazi M., "*Z-Shaped Dielectric Resonator Antenna for Ultrawideband Applications*" , IEEE Transactions on Antennas and Propagation, Vol. 58, No. 12, Pp. 4059 - 4062, 2010.

[35]   Rajveer S. Yaduvanshi, Harish Parthasarathy, "*Rectangular Dielectric Resonator Antennas Theory and Design*", Chapter 2: *Rectangular DRA Resonant Modes and Sources*, Springer India, 2016.

[36]   G. P. Junker, A. A. Kishk, A.W. Glisson and D. Kajfez, "*Effect of an air gap around the coaxial probe exciting a cylindrical dielectric resonator antennas*", Electronics Letters, Vol. 30, No. 3, pp. 177-178, 3 Feb. 1994.

[37]   G.P. Junker, A.A. Kishk, A.W. Glisson and D. Kajfez, "*Effect of air gap on cylindrical dielectric resonator antennas operating in $TM01$ mode*", Electronics Letters, Vol. 30, No. 2, Pp. 97-98, 20 Jan. 1994.

[38]   Kwok-Wa Leung, Kwai-Man Luk, Lai, K.Y.A.; Deyun Lin, "*Theory and experiment of an aperture-coupled hemispherical dielectric resonator antenna,*" Antennas and Propagation, IEEE Transactions on , vol.43, no.11, pp.1192-1198, Nov. 1995.

[39]   A. A. Kishk, A. Ittipiboon, Y. Antar, M. Cuhaci "*Slot Excitation of the dielectric disk radiator*", IEEE Transactions on Antennas and propagation, vol. 43, No. 2, pp.198-201, Feb. 1993.

[40]   R. K. Mongia, A. Ittipiboon, "*Theoretical And Experimental Investigations on Rectangular Dielectric Resonator Antenna*", IEEE Transactions on Antennas and Propagation, Vol. 45, No. 9, Pp. 1348-1356, September 1997.

[41]   D. Drossos, Z. Wu, L. E. Davis, "*Theoretical and experimental investigation of cylindrical Dielectric Resonator Antennas*", Microwave and Optical Technology Letters, Vol. 13, No. 3, pp. 119-123, October 1996.

[42]   R. K. Mongia and P. Bhartia, "*Dielectric Resonator Antennas - A review and General Design Relations for resonant Frequency and Bandwidth*", International Journal of Microwave and Millimeter-wave Computer-Aided Engineering, Vol. 4, No. 3, pp. 230-247, Mar. 1994.

[43]   These of PhD, Fauzi O. M. Elmegri, "*Model and design of small compact dielectric resonator and printed antennas for wireless communications applications*", University of Bradford, 2015.

[44]   K. M. Luk and K. W. Leung, "*Antennas series, Dielectric Resonator Antennas*", Chapter 2, "Rectangular Dielectric Resonator Antennas", Electronic & electrical engineering research studies;

[45] J. Van Bladel, "*On the Resonances of a Dielectric Resonator of Very High Permittivity*", Microwave Theory and Techniques, IEEE Transactions on, vol. 23, pp. 199-208, 1975.

[46] R. K. Mongia and P. Bhartia, "*Dielectric Resonator Antennas - A Review and General Design Relations for Resonant Frequency and Bandwidth*", International Journal of Microwave and Millimeter-Wave Computer-Aided Engineering, vol. 4, pp. 230-247, 1994.

[47] R. Kumar Mongia and A. Ittipiboon, "*Theoretical and experimental investigations on rectangular dielectric resonator antennas*", Antennas and Propagation, IEEE Transactions on, vol. 45, pp. 1348-1356, 1997.

[48] These de Doctorat en Electronique, Hedi RAGAD, '*Etude et conception de nouvelles topologies d'antennes a resonateur dielectrique dans les bandes UHF et SHF*', 22 Nov. 2013, Tunisia.

[49] Memoire de maitrise en ingenierie, Abderrahmane Agouzoul, '*Conception et realisation d'une antenne a resonateur Dielectrique à 60 GHz pour les applications souterraines* ', quebec- Canada, Aout 2013.

[50] Qinghua Lai, Georgios Almpanis, Christophe Fumeaux, Hansruedi Benedickter, Ruediger Vahldieck, "*Comparison of the Radiation Efficiency for the Dielectric Resonator Antenna and the Microstrip Antenna at Ka Band*", IEEE Transactions on Antennas and Propagation, Vol. 56, No. 11, Pp. 3589 - 3592, Nov. 2008.

[51] Debatosh Guha, Chandrakanta Kumar, "*Microstrip Patch versus Dielectric Resonator Antenna Bearing All Commonly Used Feeds: An experimental study to choose the right element*", IEEE Antennas and Propagation Magazine, Vol. 58, No. 1, Pp. 45 - 55, Feb. 2016.

[52] Odile Picon et al, '*Les antennes: theorie, conception et application*', Chapter 6: *Differents domaines d'utilisation des antennes*, Dunod, Paris, 2009.

Chapter III

Miniaturisation techniques for planar antennas

1. Introduction

The world of personal communications is developing rapidly due to the progression of RF component technology. Circuit simplicity and requirements for less congestion and lower volume, weight and cost are always in demand by manufacturers. Miniaturisation is one Ficon to meet these requirements.

This chapter looks at the various miniaturisation techniques. For each method, we will cite the latest research available in international journals.

In the course of the study of these techniques, we will give more details and examples of the first two techniques, as they correspond to our research work which will be described in chapters (IV and V).

2. Definition of electrically small antennas [1]

An Electrically Small Antenna (ESA) is defined using four types of antenna:

Firstly, in terms of their size; secondly, as a function of frequency of operation; thirdly, in relation to their use; and fourthly, to the structure constrained by size.

Basically, there are two types of ESA: an electrical element and a magnetic element. The electrical element coupled to the electric field is known as a capacitive antenna. The magnetic element coupled to the magnetic field is known as an inductive antenna [2].

Today, small antennas are mainly used for mobile communication and other wireless systems. These wireless systems are used for communication, control, detection, proximity communication, including necessary radio frequency identification, medical use, communication and body data, and video [3].

Figure Ш.1: Changes in mobile phones in recent years.

3. Miniaturisation factor [4]

The definition of antenna miniaturisation consists of reducing the overall dimensions of the antenna while retaining its key characteristics such as impëdance and radiation patterns.

For narrow-band antennas, this means obtaining a resonance with physical dimensions much smaller than the half-wavelength of the free space (λ_0) at resonance. This factor is determined by (3.1):

$$FM = \frac{f_{r\acute{e}f}^{\,original}}{f_{r\acute{e}f}^{\,miniaturis\acute{e}e}} \qquad (3.1)$$

This definition allows users to choose the reference frequency according to the application. The greater the miniaturisation factor, the greater the degree of miniaturisation.

4. Effect of miniaturisation on antenna parameters [4].

To provide a basis for the study of small antennas, an overview of the effect of miniaturisation on their most important characteristics is given below.

4.1. Director

Theoretically, it is often stated that small antennas have a unidirectional and bidirectional radiation pattern, with a directivity D ranging from 1.5 to 3. We can state that the antennas have a significant radiation in spherical mode. Small antennas are also classified as super-directional antennas, since for decreasing ka size, their directivity D remains constant [5, 6].

4.2. Radiation efficiency

Radiation efficiency is a critical issue for small antennas but has not been studied rigorously. The antenna radiation efficiency factor i] is simply the ratio of the power radiated by the antenna to the power delivered to the input terminals of the antenna.

Often the efficiency factor is represented in the formula:

$$G = \eta(1-|\Gamma|^{2})D \qquad (3.2)$$

Or ; G: realise gain.

And the efficiency of radiation i] can be represented as:

$$\eta = \frac{R_{rad}}{R_{rad}+R_{loss}} = \frac{R_{rad}}{R_{A}} \qquad (3.3)$$

Or; $R_A = R_{rad} + R_{loss}$: total antenna input resistance. The losses in the antenna, with the exception of radiation, are modëlisëes through a series loss resistance (R_{loss}).

It can be seen that as the antenna size ka decreases, R_{rad} decreases and the sëerial loss resistance R_{loss} dominates the efficiency expression in equation (3.3). Harrington [7] shows that losses are extremely large for smaller ka values.

This reduction in efficiency is mainly due to frequency-dependent conduction and diëlectrical losses in the antenna.

4.3. Antenna quality factor

A notable intrinsëque quantum for a small antenna is the Q factor, dëfined in [7]:

$$Q = \frac{2\omega_0 \max(W_E, W_M)}{P_A} \tag{3.4}$$

Or ; W_E and W_M: stored electrical and magnetic energies averaged over time.

P_A: power received by the antenna.

The power radiated is linked to the power received through it:

$$P_{rad} = \eta P_A \tag{3.5}$$

Or; η: antenna efficiency.

Another important characteristic of Q is that it is inversely proportional to the bandwidth of the antenna. A commonly used approximation between Q and the 3 dB fractional bandwidth BW of the antenna is:

$$Q \approx \frac{1}{BW} \quad \text{pour: } Q \gg 1 \tag{3.6}$$

4.4. Input impedance and correspondence

The input impedance of small antennas is generally characterised by low resistance and high reactance [8]. As the size of the antenna decreases, the radiation resistance R_{rad} decreases, causing the reactance of the X_A antenna to dominate.

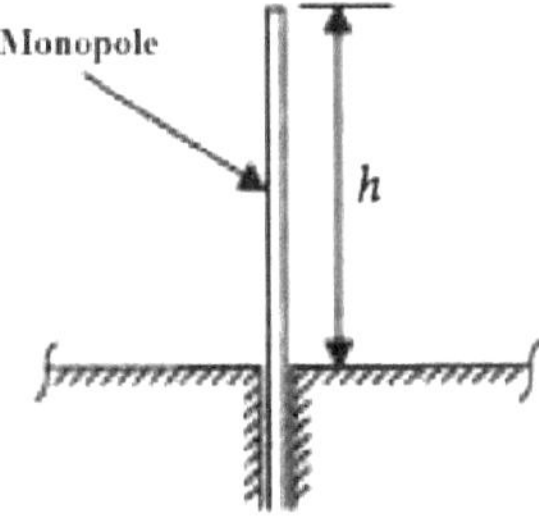

Figure III.2: Monopole antenna.

With regard to Figure III.2, a practical proportional^ relationship, for small monopole antennas, has ëlë doniK'e by [8] and [9]:

$$R_{rad} \propto \left(\frac{h}{\lambda}\right)^2 \tag{3.7}$$

Or ; h: height of the monopole.

λ: wavelength.

This implies that the input resistance decreases quadratically with size ë^C^^.

5. Theoretical limits to the miniaturisation of planar antennas [1].

A fundamental limit of the ESA has ë1.ë made by Wheeler since 1947 [2, 10, 11].

Wheeler considers the maximum dimension of the ESA to be inertial a: X/2л.

$$ka < 1 \qquad (3.8)$$

$$k = \frac{2\pi}{\lambda} \qquad (rd/m) \qquad (3.9)$$

Or ; λ: free space wavelength (metres)

a: radius of the sphere containing the maximum size of the antennae (metres).

Wheeler states that the ESAs are in free space and can be surrounded by a sphere of radius a, with: ka <1

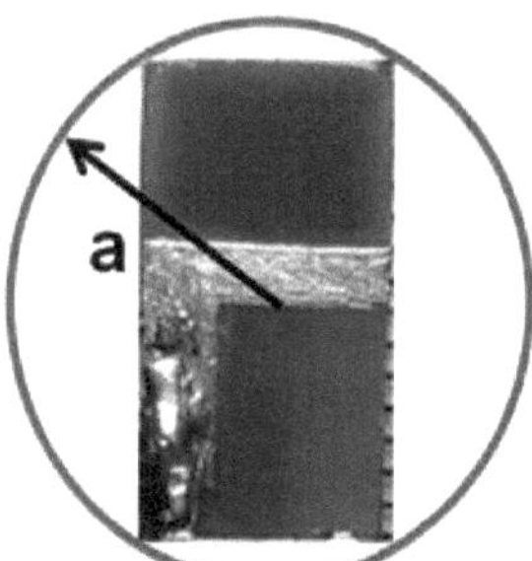

Figure III.3: Schi sphere.

Since 1987 [2, 10, 11], Fujimoto, Henderson, Hirasawa and James have also studied the theoretical limits of the ESA. They designed antennas with a minimum quality factor of Q . If the higher quality factor (Q) is lower, the impedance bandwidth is smaller.

In 1996, McLean corrected the work on Q minimum (quality factor) by one ESA.

If minimum Q for an ESA in free space:

$$Q_L = \frac{1}{(ka)^3} + \frac{1}{ka} \qquad (3.10)$$

The minimum Q for an ESA in circular polarisation:

$$Q_{L(CP)} = \frac{1}{2(ka)^3} + \frac{1}{ka} \qquad (3.11)$$

As can be seen from equation (3.10), decreasing the size of an antenna leads to an increase in its Q factor. A more realistic measure of antenna performance is the quality factor (Q) divided by the antenna efficiency (q_{ra} d). In addition, the antenna's Q factor can be reduced at the expense of its efficiency and gain (increasing losses widen the bandwidth). The design of small antennas is therefore an art of compromise between size, bandwidth and gain. Thus, once the antenna is miniaturised, there is little room to improve its bandwidth or gain. [12] Reducing the size of an antenna causes dëgradations in antenna gain and efficiency, and the bandwidth becomes narrower. [13]

The parameters affecting the fundamental limitations of small antennas are as follows:

- **Antenna quality factor Q:** Q is proportional to (ka) $.^{-3}$
- **The bandwidth ratio** (RBW): RBW is proportional to $(ka)^3$, RBW is almost ëgal

to 1/Q. q is proportional with (ka)₄.

-      **The size of the antenna ka:** ka is much smaller than the unit, where a is the radius of a sphere surrounding the antenna [14].

6. Different miniaturisation techniques

In this section, we will present in detail, the mëthods of miniaturization foundë in the literature and discuss the effect of each mëthod on the antenna characteristics.

6.1. Technique no. 1: Modification of the ground plan (DGS)

6.1.1. Description of the technique

The Defected Ground Structure (DGS) is a severe periodic or non-periodic configuration defect in a ground plane [15]. The principle of the DGS method is to use slots to modify the current distribution on the ground plane, due to the size of the DGS and its relative position, resulting in the alteration of the transmission characteristics.

The DGS has different characteristics from their ground plane geometries. It has been widely used as various microwave filters such as low-pass, band-pass and notch filters [16-18]. DGS has been used as the main method to reduce the resonant frequency for different antenna configurations in our work.

6.1.2. Work carried out and discussion

a) Various forms of DGS structure

An extensive study has been made on rectangular microstrip antennas; two different DGS structures of various slots such as a square slot, a triangular slot (see Fig. III.4) in the ground plane are discussed in [19]. It is observed that the problems encountered when making slots on the radiating patch of the antenna are also solved by using the DGS technique.

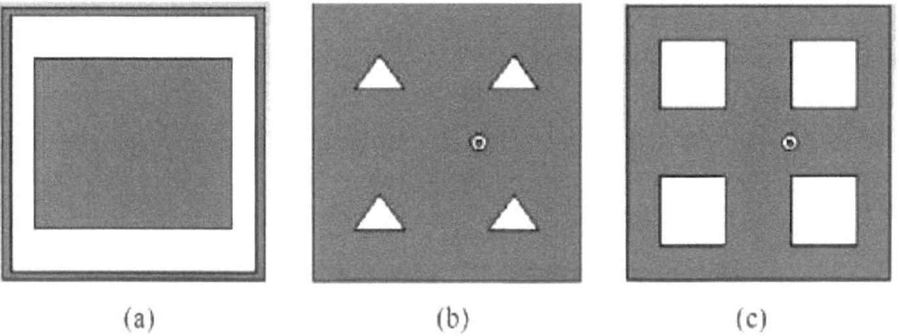

(a)   (b)   (c)

Figure III.4: (a) Front view of the patch
Rear view of patch with slit: (b) triangular and ; (c) square.

The antennas have a small size using the DGS structure for multiband applications; wireless local area network (WLAN-5.8 GHz), i'interop6rabiiit£ mondiale pour l'acces hyperfrequences (WiMAX-5.8 GHz), aero port monitoring (2.7-2.9 GHz), and direct broadcast satellite (DBS) Europe (10.7-12.75 GHz) are prësentës in this work.

These structures reduce the cross-polarized (XP) radiation field without affecting the dominant-mode input impëdance and co-polarizedës radiation patterns of the conventional antenna. The proposed antennas have shown good radiation characteristics in various operating bands, making them adaptëable to multiband opërations.

51

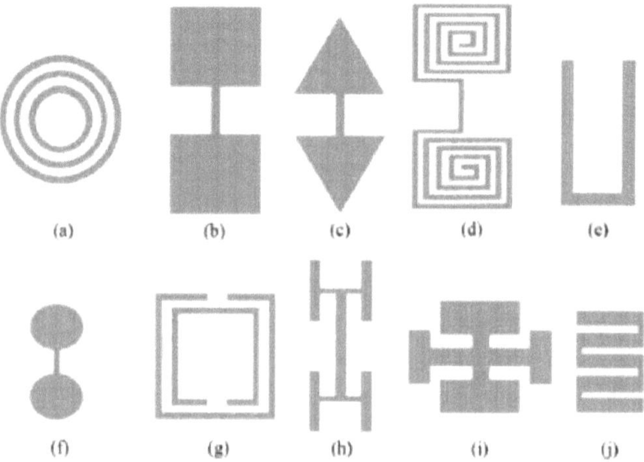

Figure Ш.5: Different forms of DGS structure. [20]

Figure III.5, shows examples of DGS structures: (a) concentric ring shaped, (b) haltere, (c) Пее1ге head haltere, (d) spiral shaped, (e) U-shaped, (f) circular head, (g) split ring Rësonators, (h) H-shaped haltere, (i) cross shape, (j) mëandre line.

b) Miniaturisation using the DGS technique

The concept of structures (DGS) has been developed to improve the characteristics of many microwave devices. To this end, DGS is also used in the microstrip antenna for certain advantages such as reducing the size of the antenna, reducing mutual coupling in antenna arrays, and so on.

In [20], the ground plane defect (GVD) structure was used to miniaturise a microstrip patch antenna and shift the resonant frequency from an initial value of 10 GHz to a final value of 3.5 GHz, without any change in the dimensions of the original microstrip patch antenna.

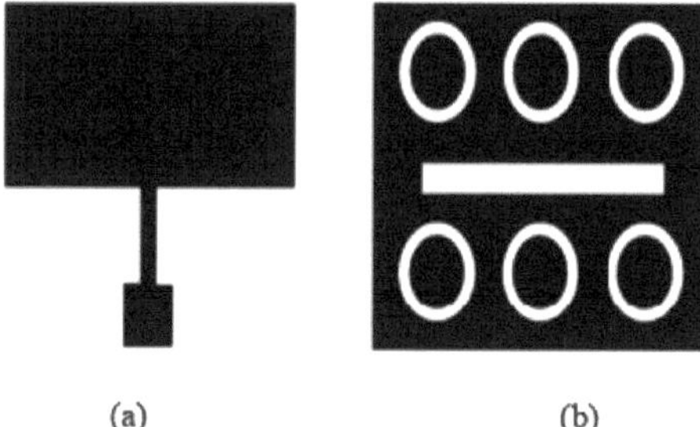

Figure Ш.6: The proposed antenna: (a) radiating element, (b) proposed DGS structure.

Figure Ш.6 shows the layout of the DGS structure, which is located on the metal ground plane. It consists of six rings with a rectangular-shaped slot; the resonance frequency can be shifted by varying the dimensions of the different concentric ring shapes or the dimensions of the rectangular slot. Geometric parameters are calculated to separate the rectangular slot from the inner and outer ring of the rings.

It can be seen that the simulated radiation pattern for the proposed antenna is bidirectional

(Figure III.7) and the maximum gain value obtained is 2.26 dB at the resonant frequency.

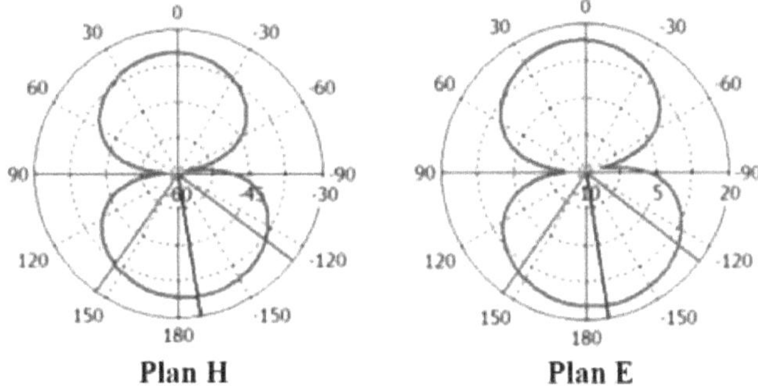

Figure III.7: 2D radiation pattern of the patch antenna at 3.5 GHz (E plane and H plane).

c) Improving bandwidth and efficiency

The microstrip patch antenna (for MPA) is used on a large scale because of its physical simplicity and low cost, but one drawback of this type of antenna is its narrow bandwidth, which sometimes limits its use in various applications.

As a solution, further research into the DGS structure has been carried out in [21], where two designs of a rectangular microstrip antenna of size 30.2x32.8x1.6 mm^3 have been proposed; one of the designs has a rectangular slot on the ground plane surface, while the other has three slots in the shape of a polygon.

The simulation results show an improvement in bandwidth and radiation efficiency after the application of the DGS structure; Table III.1 summarises the parameters of the three antenna designs.

Table III.1: Antenna parameters with and without DGS.

N°	Parameters	MSA simple	MSA with rectangular DGS	MSA with polygon-shaped DGS
1	Bandwidth	87 MHz	165 MHz	174.8 MHz
2	Reflection coefficient	-21 dB	-31 dB	-33 dB
3	Resonance frequency	3.5 GHz	3.51 GHz	3.7 GHz
4	Antenna efficiency	5 (3.33 GHz)	14.5 (3.44 GHz)	15.8 (3.26 GHz)

These two antennas have been designed to operate in the WiMax frequency band and can therefore be used for industrial, military and cellular applications.

d) DGS fractal structure

A new technique for designing a single-fed circularly polarized (CP) microstrip antenna is proposed in [22]. The CP radiation is obtained by adjusting the size of the fractal DGS structure (FDGS) etched in the ground plane.

The CP microstrip antennas with the second and third iterative FDGS are shown in Figure III.8.

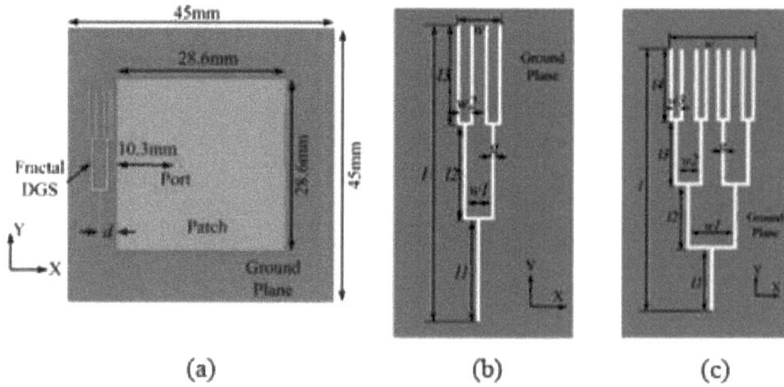

(a)   (b)   (c)

Figure Ш.8: The gëomëtries of the proposed microstrip antenna, (a) Radiating L^riment,
(b)
FDGS with a second iterative and, (c) FDGS with a third iterative. [22]

The bandwidth of the antenna (for s_{11} measured at -10 dB) is approximately 30 MHz (from 1.558 to 1.588 GHz). The gains are relatively stable in the CP measurement band, with values of between 1.7 and 2.2 dBi.

In 1 article in reference [23], a new multi-band antenna has been designed and built by combining a fractal shape and a ground plane fault structure (GFS) to achieve performance superior to gëostationary satellite communications.

The fractal DGS shape gravëe on the ground plane of the antenna has ëlë obtained in ПёпиП three Apollonius structures nested in circles to be able to tune three different bands (Figure Ш.13).

The circles of Apollonius have ëlë been used to design a serious fractal shape.

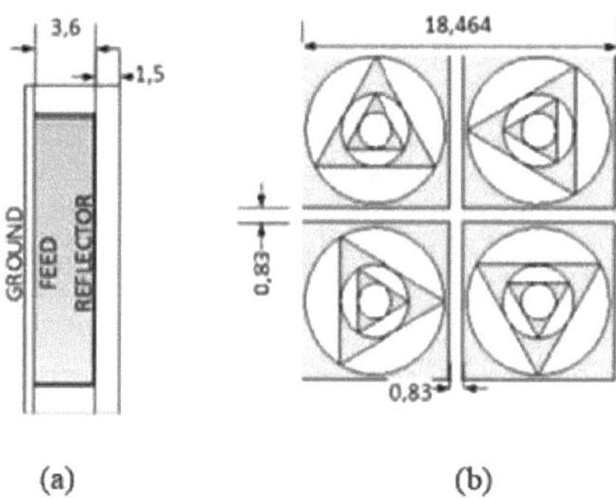

(a)   (b)

Figure III.13: Dimensions of the 2*2 array antenna in mm; (a) side viewë and (b) front view.

The race for more compact and efficient antennas is ^vitaHe for various applications such as satellite communications. Test results of the antenna prësentëe provedë its efficacyë through a more ëкyë gain and a smaller size (18.46*18.46 mm²) compared to similar attempts in the НкёгаШге.

Figure III.14 shows photos of the etched prototype.

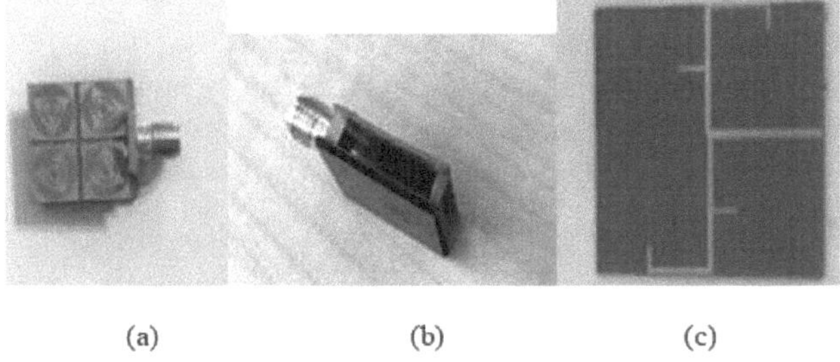

(a)   (b)   (c)

Figure Ш.14: (a) Ground plan of the 2*2 array antenna; (b) c6të view; and (c) front view.

e) Antenna matrix with DGS

A new 4 x 4 quad patch antenna array (Figure Ш.9) operating in the 28-38 GHz frequency range for 5G mobile networks is presented in [24]. To improve the radiation characteristics of the array, a ground plane fault (GFD) structure is used.

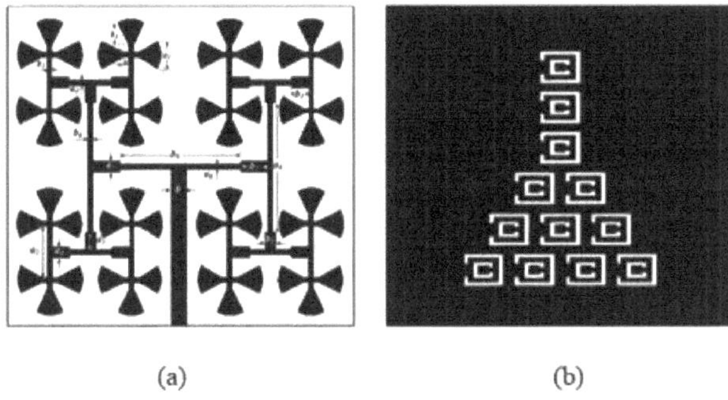

(a)   (b)

Figure III.9: Patch 4x4 antenna array (a) top view, (b) DGS (bottom view). [24]

The results show the effect of the DGS on the antenna's performance characteristics:

- Radiation efficiency and the reflection coefficient are improved, on average, by around 17.14 and 69.2% respectively.

- The actual antenna gain was increased by 2.44 dBi with a negligible effect on the beamwidth at half-power.

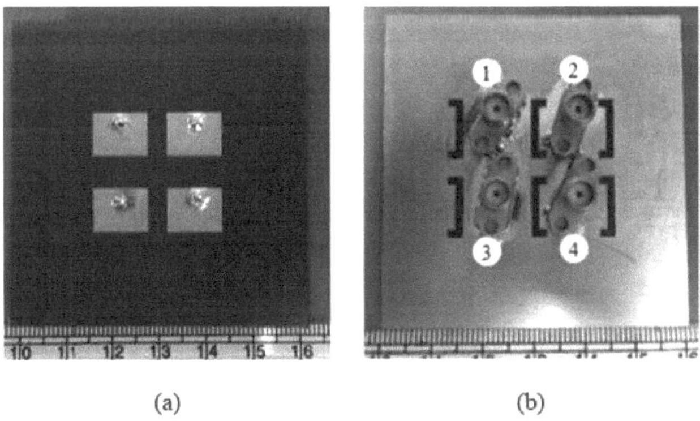

(a)   (b)

Figure III.11: Photos of the 2x2 matrix with the DGS structure;

(a) from the patch side, (b) from the ground plan side. [25]

Further work of the integrated microstrip array of the DGS structure has ëlë explored and dëmontrë in [25], the main aim of which is to achieve improved polarisation purity. A DGS formed with the variable configuration has been designed (Figure III. 11) for X-band applications, showing a 12 dB improvement in isolation between main and cross-polarised radiation.

This improvement, which has been verified experimentally, allows the XP to be reduced without affecting the primary radiation or gain values (the peak gain being 12.2 dBi). The considerable uniformity of the substrate fields obtained by DGS can be seen from Figure III.12.

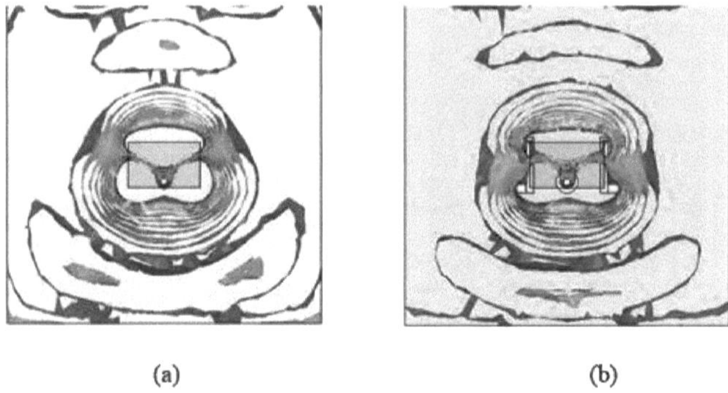

(a)   (b)

Figure III.12: Simulated substrate electric field (a) without DGS, (b) with DGS.

f) DGS structure and form in meandre

In the phased array in paper [26], the radiator sharing approach was implemented using a mëta surface antenna, and the scanning range a ël.ë amëliorëe by applying meandering grooves cut into the ground plane, forming DGS structures (Figure III.10).

The nine-element network with DGS achieves a measured impedance bandwidth of 23% (4.6 ~ 5.8 GHz) and the realised gain ranges from 14.76 to 11.85 dBi.

Based on the good directivities of the proposed antenna element, the phased array presents high gains and narrow scanning beams. It may therefore be a promising candidate for high-resolution radar systems.

On the basis of theoretical analysis and experimental simulation, it is shown that the concept of using periodic metamaterial antennas with a radiator-sharing approach enables the design of wide-angle scanning phased arrays.

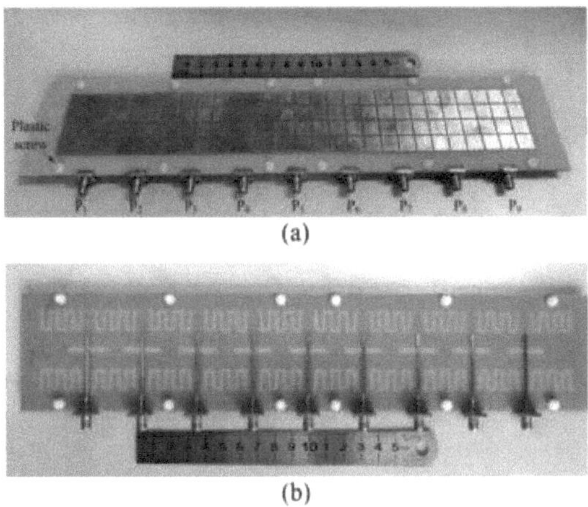

(a)

(b)

Figure Ш.10: Photos of the iabricated phase matrix with DGS.

(a) Top view and (b) bitter view. [26]

g) DGS structure with dielectric resonator antennas (DRA)

A novel two-element fractal tree DRA matrix design for wideband multiple-output multiple-input (MIMO) applications is proposed in [27].

A fractal tree geometry is proposed (Figure III.15) to achieve broadband characteristics while a periodic C-shaped DGS structure (PDGS) is used to achieve a reduction in mutual coupling between two closely spaced antenna elements without unduly affecting the bandwidth of the DRA network.

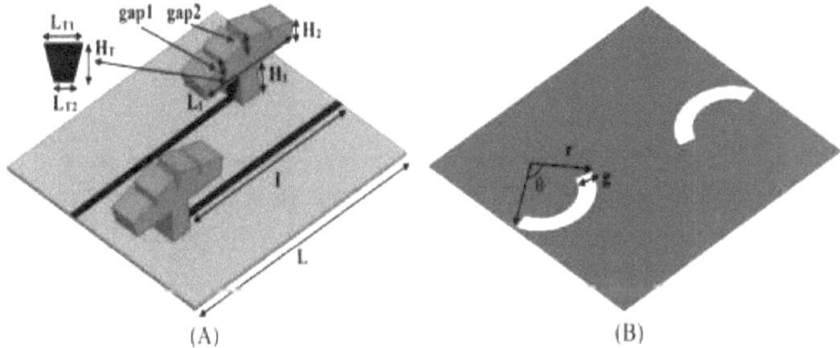

Figure III.15: Gëomëtrie of the DRA design has two fractal tree ëlëments:

(A) front view, and (B) back view.

Figure Ш.16 shows the surface current distributions on the antenna ground plane before and after slot insertion (PDGS).

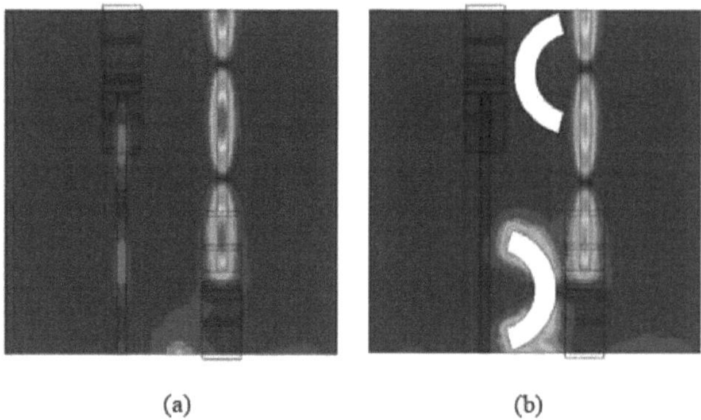

(a)   (b)

Figure III.16: Surface current distribution on the ground plane ; (a) of the antenna (without PDGS) and ; (b) of a tree-shaped

DRA array with PDGS (proposëe design).

The following conclusions are drawn:

- Bandwidth can be improved by fractal notches when they are introduced at the right positions.

- The përiodic C-shaped DGS structure helps to reduce mutual coupling to below 21.5 dB across the entire band of intërët.

- A measured impëdance bandwidth of approximately 89.9% from 3.95 to 10.4 GHz has been achieved. In addition, the mutual coupling performance measured was less than 21.5 dB over the entire interference band.

- A similar radiation pattern across the entire interest band was achieved, with around 98% radiation efficiency.

- In terms of broadband characteristics and mutual coupling performance, the proposed antenna may be a good option for MIMO applications.

h) DGS structure and reconfigurable antennas

In reference [28], a reconfigurable semi-transparent frequency antenna is proposed and ëtudiëe. This antenna is implemented on glass as a substrate, AgHT-4 as a radiative element and copper as a ground plane to achieve a semi-transparent characteristic. A pair of PIN diodes are used as microwave switches with a DC bias circuit.

This proposed antenna is fed by a CPW device and introduces an E-shaped DGS structure on the ground plane (Figure III. 17) for electrical length modification. The antenna has a -10 dB bandwidth from 3 to 6 GHz when the PIN diodes are on, and reconfiguration to a narrow-band centre mode at 4.75 GHz when the PIN diodes are off.

58

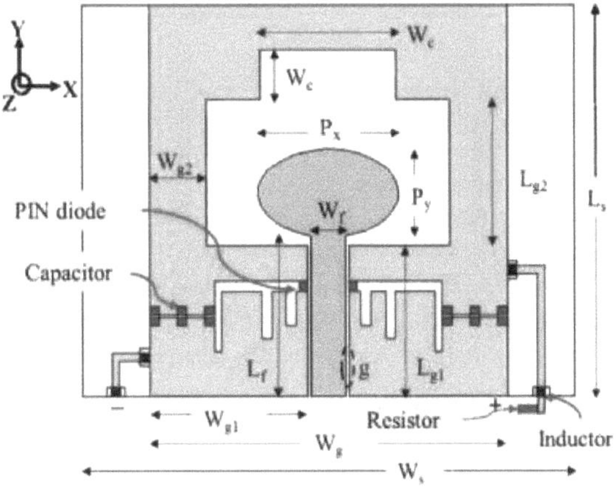

Figure III.17: Geometry and DC bias circuit of the proposed antenna.

The evaluation results showed good agreement between simulations and measurements, with quasi-omnidirectional patterns in the xz plane and bidirectional radiation in the yz plane.

A new antenna structure comprising a DGS structure with a hook-shaped radiating patch designed for wide bandwidth is presented in [29]. A small rectangular slot is removed from the ground plane just below the radiating patch for better impedance matching. The ground plane is obtained by etching 3 straight symmetrical slots to miniaturise the antenna (Figure III.18).

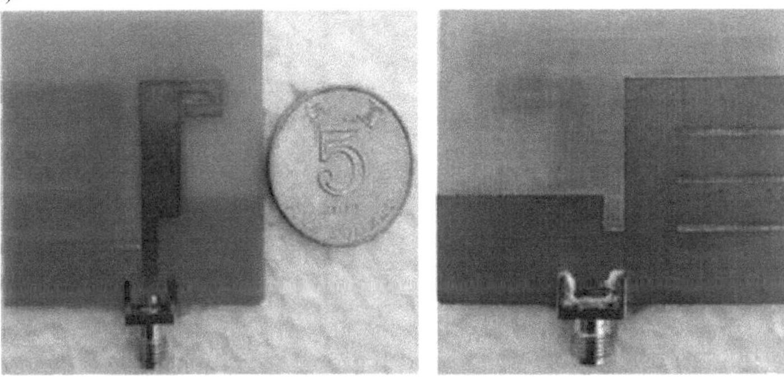

Figure III.18: Top and bottom of the fabricated antenna. [29]

The results obtained show that the antenna can produce a bandwidth of around 27.60% centred on a frequency of 2.17 GHz. The maximum measured gain is 4.30 dBic, while more than 90% of the radiation efficiency is achieved throughout the CP radiation bandwidth.

6.2. Technique n°2: Loading with very high permittivity materials

6.2.1. Description of the technique

High-permittivity materials allow good miniaturisation of microwave components, but the antenna using these materials has the disadvantage of a reduced bandwidth.

A high permittivity dielectric can be used:

-        by loading it onto the substrate, to build a dielectric resonator antenna (DRA) with high permittivity.

-        or as a substrate to obtain a very small resonator.

These two types are useful for making a compact antenna for an applicable wireless communication system by reducing the need for large space as well as the production cost.

6.2.2. Work carried out and discussion

a) Dielectric resonator antenna (DRA)

The widespread use of wireless communication systems has prompted researchers to develop small dielectric resonator antennas (DRAs), the aim of which is to have no conduction loss and effective coupling with almost all transmission lines [30]. The physical size of a DRA is proportional to l/^/s^, where; m is the relative permittivity of the dielectric resonator (DR). As the relative permittivity increases, there is a natural tendency to reduce the size of the DRA. However, a high permittivity DRA offers a narrow bandwidth that limits its immediate practical applications. Hence the development of a low-profile DRA with improved bandwidth is essential in this context.

Several low-profile DRAs suitable for microwave communications have already been reported:

Paper [31] deals with the incorporation of a high permittivity cylindrical dielectric resonator (CDR) into a circular microstrip antenna with planar microstrip lines as the feed line. The RDC used has a relative permittivity of 50 with a radius of 6.75 mm and a height of 6.7 mm. The microstrip antenna is placed on a ëpoxy dielectric substrate (FR4) with a relative permittiv^ of 4.2 and a ëthickness of 1.6 mm (Figure III.19).

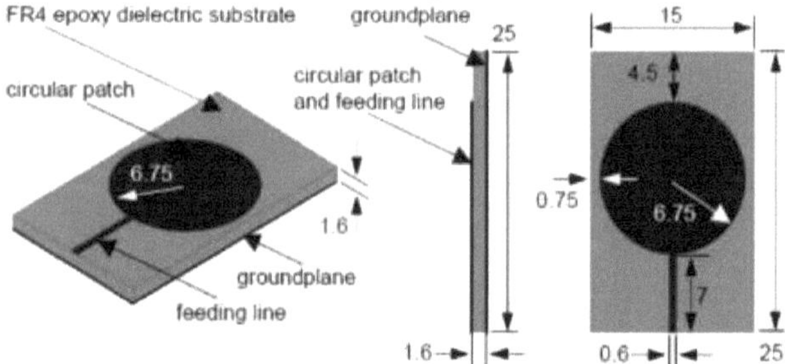

Figure Ш.19: The microstrip antenna without RDC (unhe in mm).

The RDC is incorporated into the microstrip antenna by folding it ташёге concentrically onto the circular patch of the microstrip antenna, as is shown in Figure III.20.

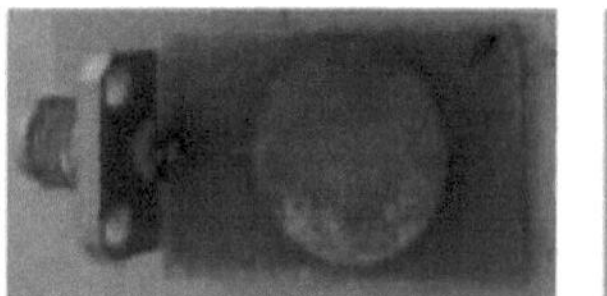
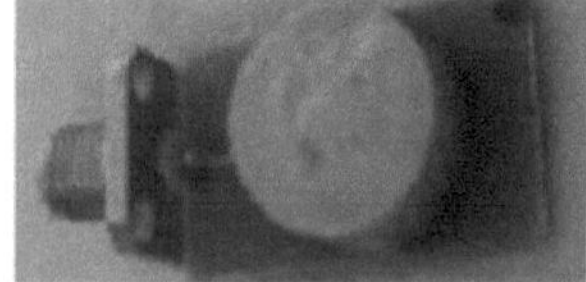

Figure III.20: Prototype of the rëa^ëe microstrip antenna without and with CDR.

The properties of the microstrip antenna with RDC were also compared with that without RDC. The experimental measurement shows that the resonance frequency of the high-permittivity circular diëlectrical resonator antenna is about 1 GHz lower than that of the microstrip antenna without RDC and shifts from 5.94 GHz to 5.04 GHz (Figure III.21). The incorporation of a high permittivity cylindrical diëlectrical resonator^ (RDC) in the microstrip antenna could lower the resonant frequency of the antenna without increasing its size.

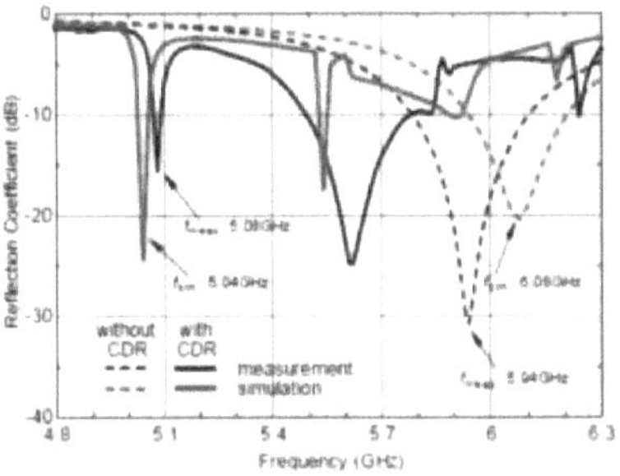

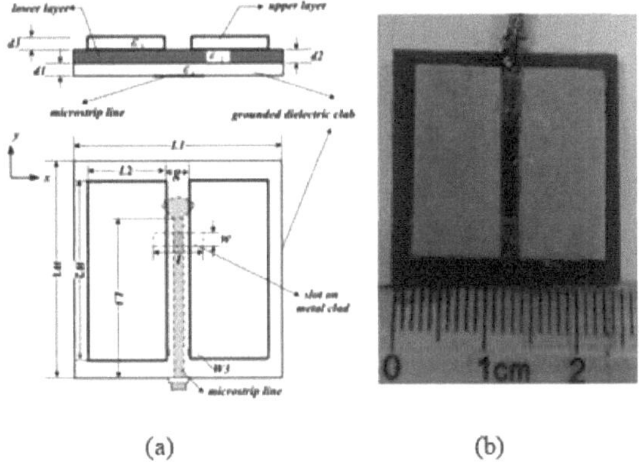

Figure III.21: Simulated and measured reflection coefficients for the two antennas. Although there were some discrepancies in terms of resonance frequency and bandwidth, in general the two antennas showed good agreement in their performance.

A very high permittivity dielectric resonator antenna has been proposed in [32]. The antenna is designed on the basis of a multilayer structure. Two separate high permittivity dielectric resonators (ra 2.2 and aɔ=93) are placed symmetrically above the antenna structure. A low permittivity dielectric substrate (ra 4.4) is inserted between the resonators and the ground plane. The overall height of the antenna is 1.57 mm (Figure III.22).

(a)   (b)

Figure III.22: (a) Geometry; and (b) Photo of the proposed antenna.

The s_{11} parameter simulated and measured in relation to frequency is shown in Figure III.23. The proposed antenna is tested at 7.5 GHz. It can be seen from the simulated results that the relative impedance bandwidth is 9.67% (7.18-7.91 GHz), in which s_{11} is less than -10 dB. The measured relative bandwidth can reach a tau of 10.49% (7.13-7.92 GHz). The slight difference between simulated and measured results may be due to measurement error and manufacturing tolerance.

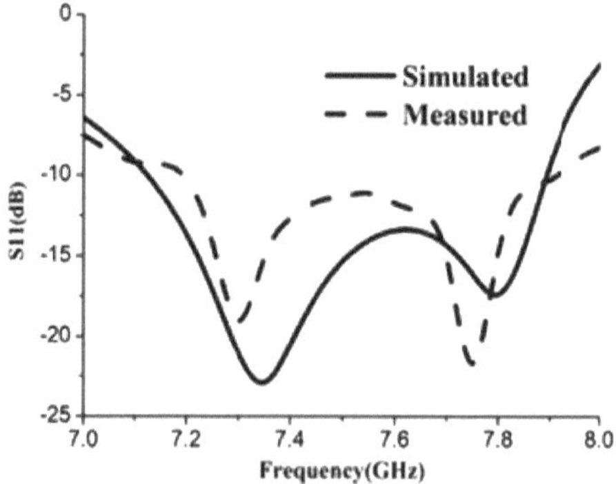

Figure III.23: Simulated and measured S11 as a function of frequency.

The measured results also indicate that stable radiation patterns throughout the operating band have been achieved. At the same time, gains in excess of 6 dBi and a low cross-polarisation level of -25 dB in the operating band were achieved.

In the rëférence [33], a ëtude that proposes a new dïélectrical resonator antenna (DRA) loaded with rectangular cëramique of the BST (Bao.8Sro.2TiOs) material.

The antenna is made up of two stacked dïélectric resonators; the first resonator is made from TMM10i dïélectric material with a permittivity of mi 9.8 and a height of D_i. The upper element is made of a thick film of BST ceramic (Bao.8Sro.2TiOs material) with a thickness of D_2 and a very high permittivity in the microwave spectrum ($\varepsilon_{r2}=250$), with: $D_i+D_2=3.5$ mm.

The entire structure is mounted on a TMM6 substrate with a constant dibelectricity of a $_{rs=6}$ and a thickness of h=0.762 mm.

The ground plane is printed on the bottom surface of the dïélectrical substrate. The excited ëléments are excited via a microstrip transmission line.

Figure III.24 shows the configuration of the propos'e structure.

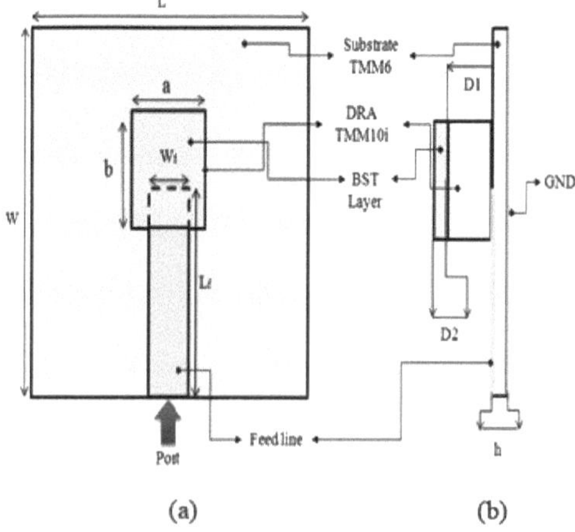

Figure 23: 24: Antenna design; (a) Top view (b) Side view.

The reflection coefficients of the antenna before and after loading the BST material are shown in Figure III.25. From these curves, it can be seen that the integration of the BST layer leads to a shift in resonance frequency from 10.7 GHz to 8 GHz (suitable for radar applications), resulting in a reduction in resonator antenna size of around 67% compared with an ordinary DRA for the same resonance frequency.

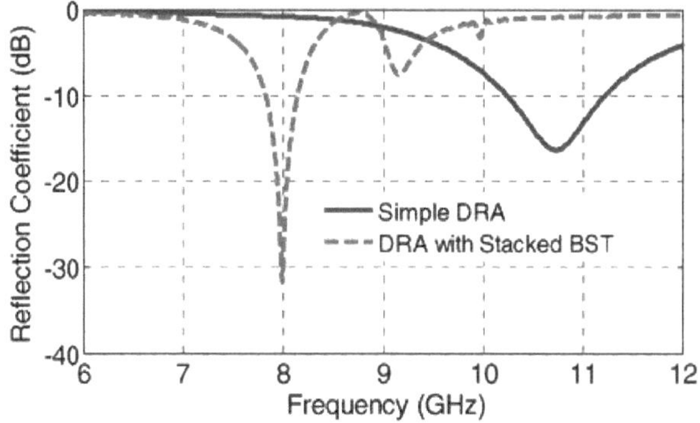

Figure 24: 25: Reflectance coefficient of the antenna with and without loading by the BST layer

The high permittivity of the matëriau allows the resonant frequency of the antenna to be shifted from 10.77 GHz to 8 GHz. In addition, the miniaturised antenna provides a high gain (6.5 dB) compared with the simple DRA (4.3 dB).

The paper [34] describes a miniaturised antenna array technology with high permittivity ceramic matëriau. The miniaturisation of planar printed circuit antennas by loading the high permittivity ($\varepsilon_r = 15$) bismuth titanate (BiT) ceramic matëriau onto the microstrip antenna has been demonstrated experimentally. Both antennas (Figure III.26) are designed using an RT/Duroid 5880 microwave substrate with a permittivity of 2.2 from Rogers Corporation.

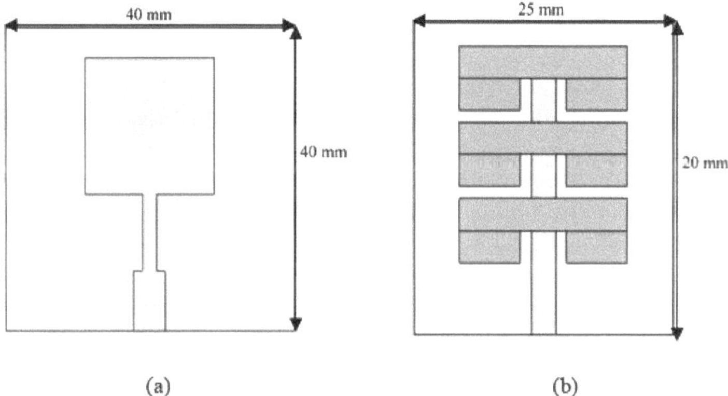

Figure III.26: Antenna design; (a) conventional microstrip antenna (b) BiT array antenna. The use of ceramic matëriau with high permittivity allows a dramatic reduction in size. The s_{11} reflection coefficient result shown in Figure III.27 indicates a resonance of around 2.4 GHz.

Conventional microstrip patch antenna BiT array antenna

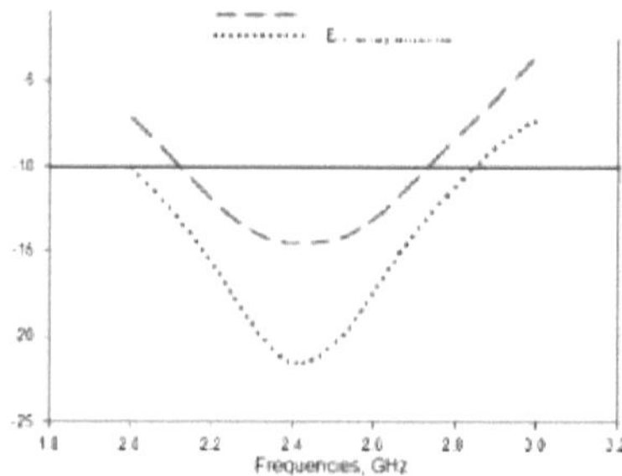

Figure III.27: S11 of the conventional microstrip antenna and the BiT array antenna.

The effects of loading ceramic material on gain, bandwidth, directivity and radiation efficiency compared with the conventional microstrip antenna are summarised in Table III.2.

Table III.2: Performance and physical size of the two antennas.

N°	parameters	Conventional microstrip antenna	BiT network antenna
1	Dimensions (mm)^3	40 x 40 x 1.6	25 x 20 x 1.6 cm
2	Gain (dBi)	3.1	7.1
3	Directivity (dBi)	4.3	7.5
4	Radiation efficiency (%)	72.1	94.7
5	Bandwidth (%)	24.2	35.4

The high radiation efficiency of the BiT network antenna is due to a lower conduction loss compared with the conventional microstrip antenna with over 90% metal base. The antenna miniaturisation achieved by using a high-permittivity ceramic material is proven and the miniaturised antenna can then be integrated with a wireless router and wireless modem.

The paper in reference [35] describes the miniaturisation of the planar inverted F antenna (PIFA) by dielectric loading of very high permittivity barium tetra-titanate ceramics at 1.8 GHz. He demonstrated that the simple loading method can reduce antenna size but at the expense of antenna performance. For example, the antenna gain becomes lower.

Next, a sophisticated loading approach based on a novel substrate-superstrate structure is developed to improve antenna gain.

The antenna is fed by a standard 50 Q coaxial probe and consists of a radiating element that has a surface area of 15.4x8.8 mm^2 , printed on a $_{Sri=38}$,80 ceramic substrate. The cë ceramic superstrate of $_{:_r}$ 2 80 with a size 49.5x49.5x4.69 mm^3 is placë above the radiating element. The 2.2 mm wide shorting plate is usedëe for grounding the radiating element and is located on the edge of the ceramic substrate so as to facilitate the manufacturing process (Figure III.28).

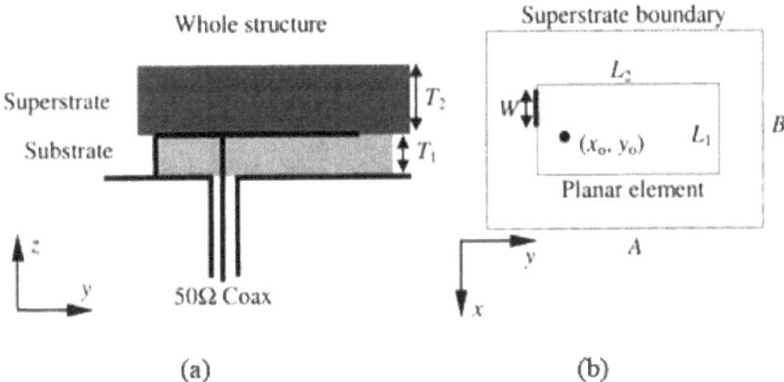

Figure III.28: Geometry of the dielectric-loaded PIFA; (a) Side view, (b) Top view.

The sophisticated loading approach basedëe on the new substrate-superstrate structure has ëlë dëveloppëe to give the 1.8 GHz antenna an amëliorëe performance with gain up to 7 dBi and bandwidth of 103 MHz (5.7%), comparable to those of conventional PIFAs.

This sophisticated loading technique can reduce the size of the antenna and maintain the same performance.

b) High permittivity substrates

State-of-the-art technology for microstrip patch antennas has led to the miniaturisation of these devices. The development of new high-permittivity substrates has enabled a considerable reduction in the physical dimensions of microstrip antennas.

Consequently, the dëvelopment of several types of diëlectric substrates, such as cëramique substrates has ëlë ^ necessary in recent dëcennies for application in many microwave devices and communications circuits.

Lately, a wide range of new applications has ëlë observed with the dëvelopment of cëramics due to their ëtës ëlectronic and diëlectric properties as well as their refractive index ëкуë and chemical stability. The paper [36] proposes the use of titanium dioxide cëramics (the diëlectric constant is ëlevëe ~ 100) as a diëlectric substrate of a microstrip antenna, with a fractal дёотё^, for applications in wireless communications. The advantages of using fractal geometries in antenna designs include miniaturisation and provide multi-band resonant frequencies.

Figure III.29 (a) shows a titanium dioxide ceramic with a diameter of 24 mm and a height of 2 mm obtained after the svntherisation process.

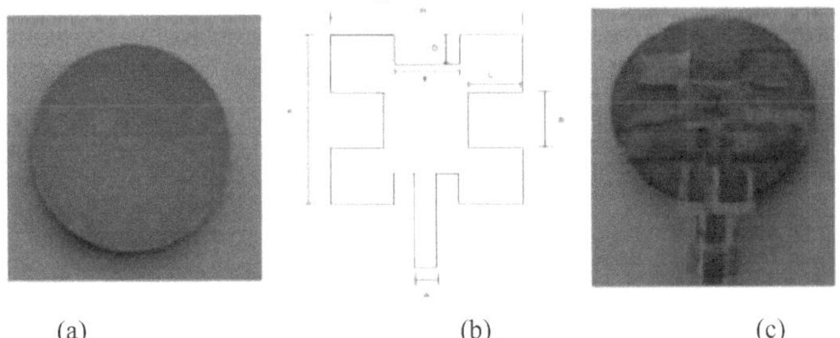

(a)   (b)   (c)

Figure III.29: (a) Ceramic titanium dioxide substrate, (b) Fractal microstrip antenna gëomëtrie, (c) Photo of the proposed fractal antenna gëomëtrie.

fractal microstrip, (c) Photo of the proposed fractal antenna gëomëtrie.

The result of measuring the s_{11} parameter between the 2 GHz and 8 GHz frequencies (Figure

III.30) shows that fractal microstrip antennas with dielectric ceramic layers are excellent candidates for the use of small devices in wireless applications.

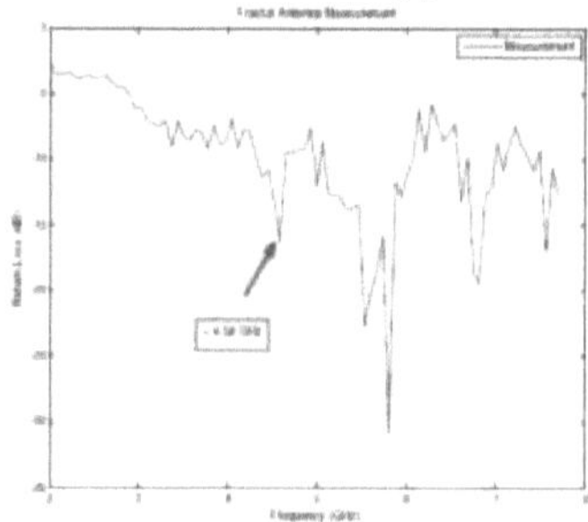

Figure III.30: Parameter s₁₁ as a function of frequency.

In paper [37], the reflection coefficient of the monopole antenna fed by a coplanar waveguide, as shown in Figure III.31(a), was analysed as a function of the relative dielectric constant on an FR4 substrate.

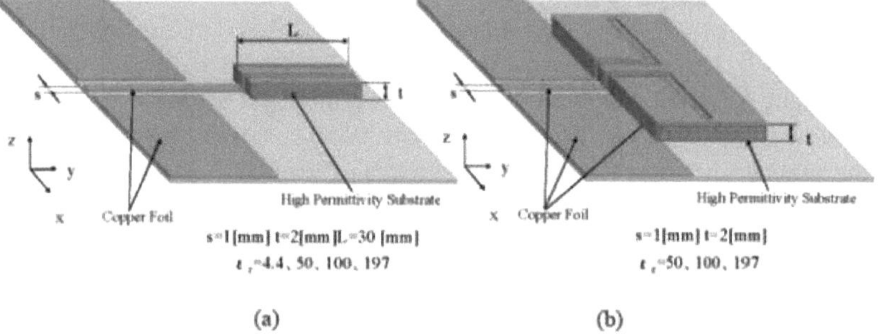

Figure III.31: (a) Printed monopole antenna, (b) Printed slot antenna etched on a diëlectrical substrate with permittivity élevëe.
diëlectrical substrate with permittivity élevëe

Figure III.32 shows an example of the calculated reflection coefficient as a function of frequency by changing the relative diëlectrical constants from 4.4 to 197. From the curves in this figure, the higher the relative dielectric constant, the lower the resonance frequency, although the bandwidth becomes narrow for use at high permittivity.

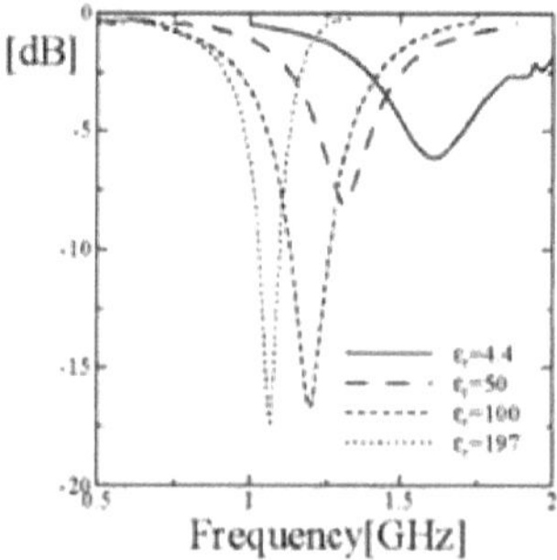

Figure III.32: Calculated ryflection coefficienty of the monopole antenna as a function of

66

the
frequency.

Based on the configuration shown in Figure III.31(b), two types of antenna with a two-slot and four-slot spiral configuration, to reduce their sizes, have been fabricated as shown in Figure III.33.

These spiral slot antennas yty etched on a high permittivity carry substrate, having a relative diëlectrical constant of 197 and a length of 10 mm^2 on each side with a ythickness of 2 mm. The width of the slit and the space between the slits have yty fixedys to 0.2 mm each.

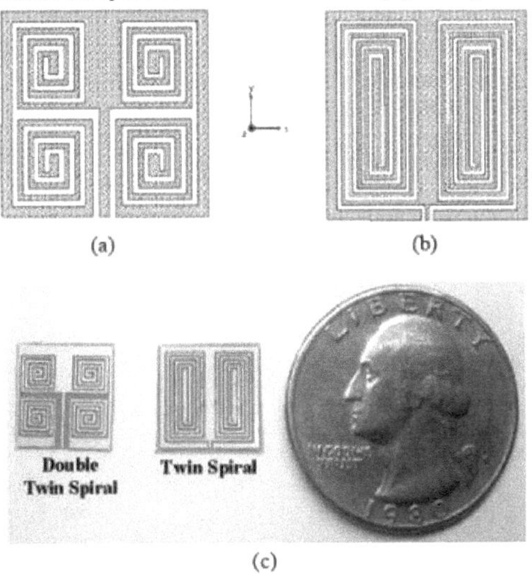

(a)   (b)

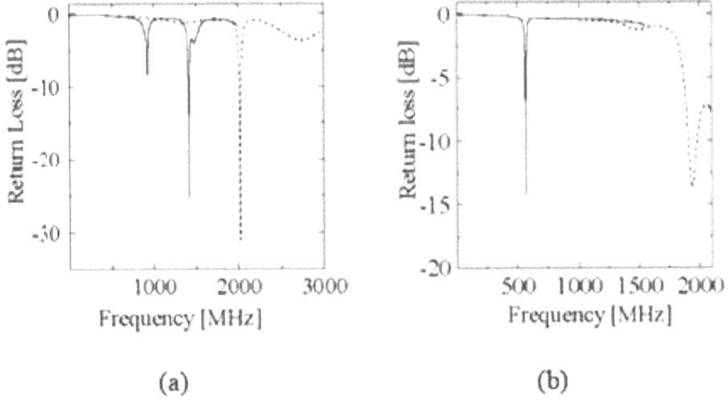

(c)

Figure III.33: Plan view of spiral antennae: (a) has four slots, and (b) has two slots.
(c) Photo of the two spiral antennae.

Initially, the four-slot spiral antenna, as shown in Figure III.33(a), was ël.ëe tested. Its measured reflection coefficient is shown in Figure III.34(a). For comparison, the resonance frequency for a high permittivity substrate became lower at 640 MHz than that for a low permittivity substrate, measured at 2000 MHz. A reasonable antenna gain of -5 dBi has ë1ë obtained.

The same applies to the double-slot spiral antenna.

(a)   (b)

Figure III.34: Measured reflection coefficients of spiral antennas: (a) with four slots, and (b) with two slots. The relative diëlectrical constants are 2.6 (dashed curve) and 197(solid curve).

6.3. Technique 3: Integration of localised elements

6.3.1. Description of the technique

Integrating active components (diode, transistor) or passive components (resistor, capacitor, inductance) between the conductive tracks (ground plane or patch) makes it possible to miniaturise microrunanium antennae.

PIN diodes are generally more widely used than transistors as switching devices for RF and microwave communication systems, because they have a number of decisive properties: good isolation, low power consumption, low insertion loss and low cost [38].

The use of ^sistive components in antennas increases ohmic losses, dëgrade the antenna's radiation efficiency and change the radiation pattern.

A low resistance value (a few ohms) is equivalent to a short-circuit (the short-circuit technique will be discussed in this chapter in section 6.4).

The use of capacitive or inductive components disturbs the field distribution of the structure and increases the ëlectrical length of the antenna, enabling miniaturisation to be achieved.

6.3.2. Work carried out and discussion

a) PIN diode (switching)

One method of achieving multiband response is to incorporate fractal geometry into the microstrip patch. Using fractal iteration on the patch antenna increases the bandwidth.

A compact reconfigurable antenna [38] with a fractal design using PIN diodes is proposed for wireless applications. The simulated frequency bandwidth of the antenna covers 3.58 GHz-8.72 GHz.

Figure III.35 shows the design of the first, second and third iterations of the reconfigurable fractal antenna controlled by six PIN diodes integrated on the structure with a 50 Q microstrip feed line. An iteration factor of 0.5 is used in this technique.

The PIN diodes are connected to the branch connection position, the links between 1[ere] iteration and 2[eme] iteration are established by diodes D1 and D2, while the link between 2eme iteration and 3eme iteration is established by diodes D3, D4, D5 and D6 (Figure III.35).

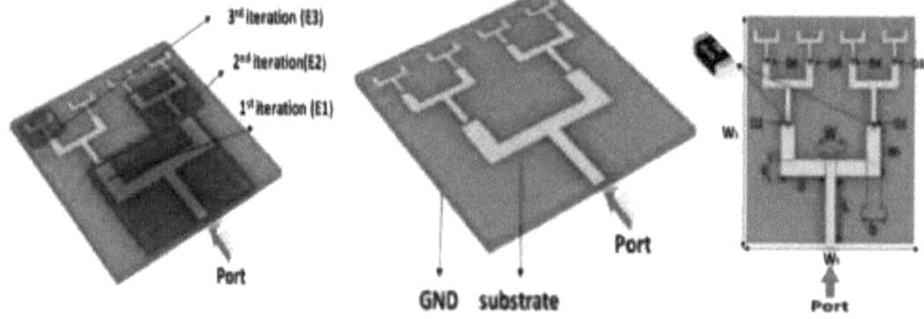

Figure III.35: Gëomëtrie and iterations of the proposed reconfigurable fractal antenna. [38] Figure III. 36 shows the reflection coefficient as a function of frequency by turning the various switches (diodes) OFF and ON.

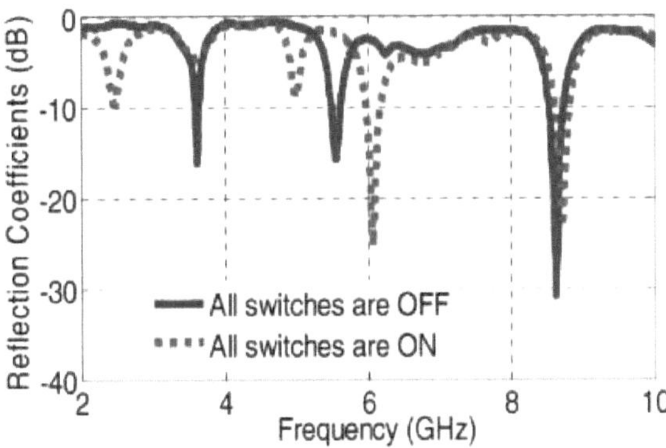

Figure III.36: Parameter s_{11} of the proposed antenna, simulated with respect to frequency. When all the diodes are in the OFF state, we can see that the antenna operates in three bands comprising; 3.62 GHz, 5.55 GHz and 8.63 GHz, but after switching all the diodes back to the ON state, we can see that the antenna operates around 2.47 GHz, 6.07 GHz and 8.72 GHz, and covers certain service bands such as: WiMAX (2.400 to 2.483) GHz, m-WiMAX (3.4 to 3.6), WLAN (5.15 GHz to 5.825 GHz), C-band (4 GHz to 8 GHz) and X-band (8 GHz to 12 GHz), which can be used for satellite and radar applications and is suitable for defence and secure communication.

In reference [39], a new frequency reconfigurable square slot microstrip antenna with an overall area of 20 x 20 mm^2 is presented. By implementing PIN diodes in the antenna structure, switchable frequency responses such as Bluetooth, WiMAX and WLAN bands are achieved. In order to cover lower frequencies, miniaturisation techniques such as ground plane modification (GMP) and the insertion of a cross-shaped sleeve have been used for Bluetooth applications.

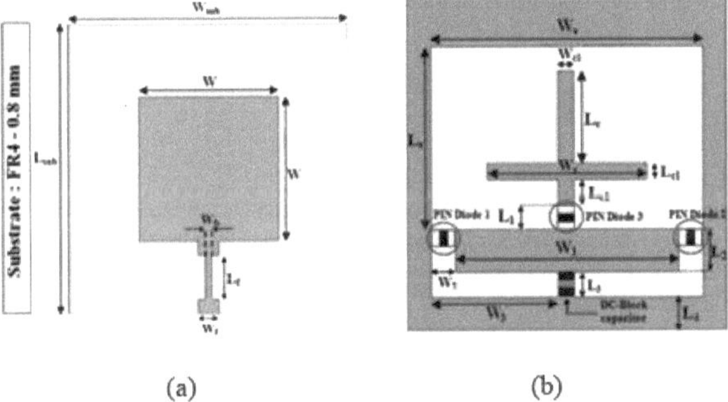

(a)　　　　　(b)

Figure III.37: Gëomëtrie of the proposed antenna: (a) top view, (b) bottom view.

In addition, the prototype of the designed antenna has ëlë iabrique (Figure III.38 shows the photo of the rëalisëe antenna). BAR64-3W PIN diodes, which have low reverse-bias capacitance (typically 0.17 pF at frequencies above 1 GHz) and low forward resistance (typically 2.1 Q at 10 mA), were used as switches.

As can be seen in Figures III.37 and III.38, a DC blocking capacitor is used in the PIN diode bias circuit to prevent short-circuiting. The DC supply is applied to the PIN diodes by means of wires. These wires can affect the performance of the antenna; spëcially the radiation pattern.

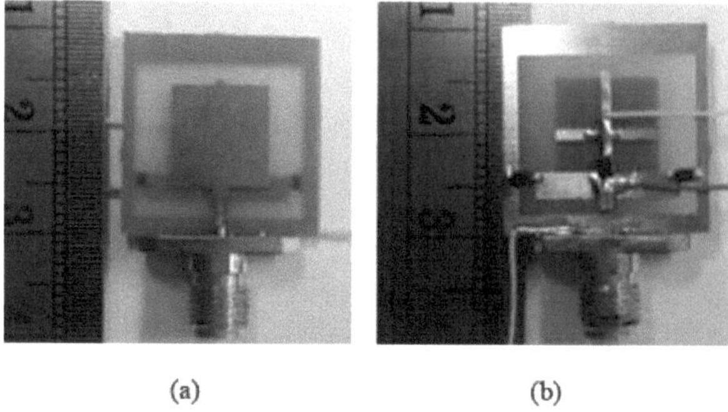

(a)   (b)

Figure III.38: Photo of the completed antenna: (a) top view, (b) bottom view

The simulated frequency reflection coefficient characteristic of the proposed reconfigurable antenna for different PIN diode bias conditions is plotted and compared in Figure III.39.

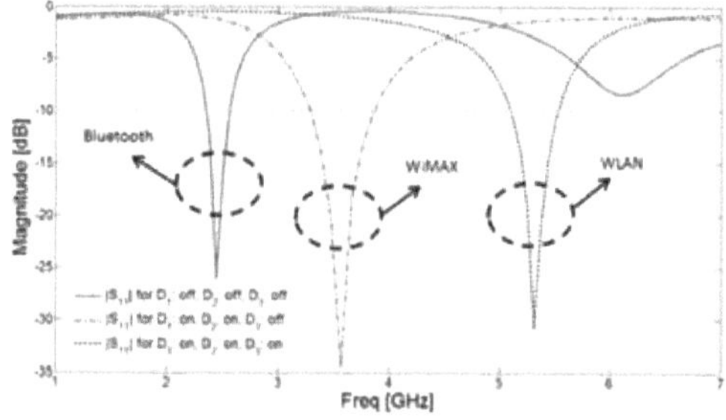

Figure III.39: The frequency responses of the proposed reconfigurable antenna for three different combinations of bias conditions for the PIN diodes implemented.

Inserting three PIN diodes into the proposed structure, as shown in Figure III.37, creates a reconfigurable antenna capable of covering Bluetooth (2.4 GHz), WiMAX (3.5 GHz) and WLAN (5.5 GHz) systems with a ratio of 8.7%, 11.2% and 11%, respectively.

b) Varicap diode (Varactor)

A compact frequency reconfigurable microstrip antenna is presented in [40]. Miniaturisation of the antenna size is achieved by a ring slot loading while frequency reconfigurability is achieved by using a varactor diode across the slot in the ground plane. The frequency agile design can be used to cover several well-known frequency bands and provides smooth frequency reconfigurability between 2 GHz and 2.3 GHz.

The geometry of the reconfigurable and miniaturised APM antenna is shown in Figure III.40. The antenna is designed on a FR4 Ur 4) substrate with a thickness of 0.8 mm and a dimension of 50*50 mm .^{2}

70

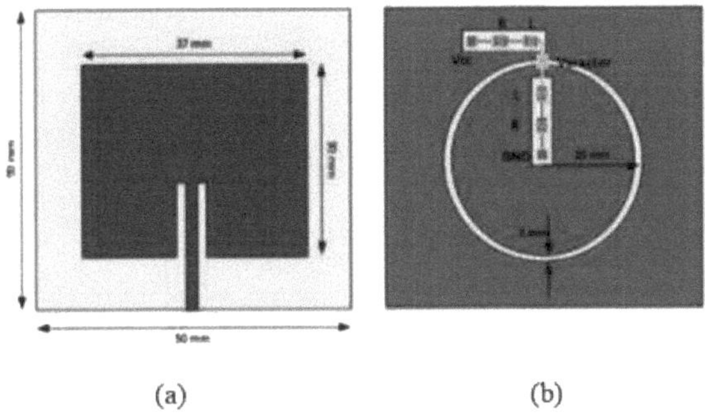

(a)   (b)

Figure III.40: Geometry of the reconfigurable and miniaturised APM, (a) Top face, (b) Bottom face

An annular slit has ë1.ë gravëe in the ground plane for miniaturisation. The slot has a radius of 15 mm and a width of 1 mm. This slot loading resulted in a decrease in the APM resonance frequency from 2.5 GHz to 1.75 GHz.

To add the reconfigurability function, a varactor diode (BB145) was fixed in the ground plane through the slot with the DC bias circuit. This provided a capacitive load to the antenna.

The variation in the varactor diode's capacitance causes a shift in the antenna's operating frequency. To this end, two additional rectangular slots have been created in the ground plane to accommodate the localised components of the bias circuit.

Figure III.40(b) shows the varactor diode with the bias circuit in the ground plane of the APM. The inductors acted as RF chokes while the resistors ë1.ëes usedëes to limit the current in the circuit.

The varactor diode has ë1ë modëlisëe as a capacitive ëlëment whose value has ë1.ë variëe. After a thorough analysis of the antenna, it ë1ë fibrillated. Figure Ш.41 shows the upper and inner côtë of the fabricated antenna.

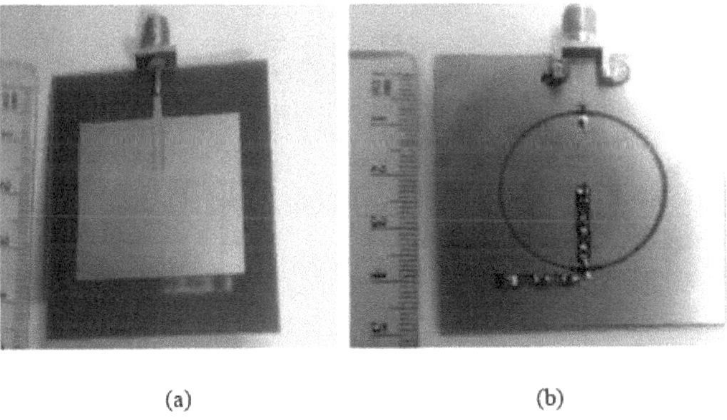

(a)   (b)

Figure III.41: The reconfigurable and miniaturised APM, (a) Top side, (b) Bottom side

Figure III.42 shows the measured reflection coefficients of the antenna obtained by varying the DC voltage across the varactor diode using the bias circuit.

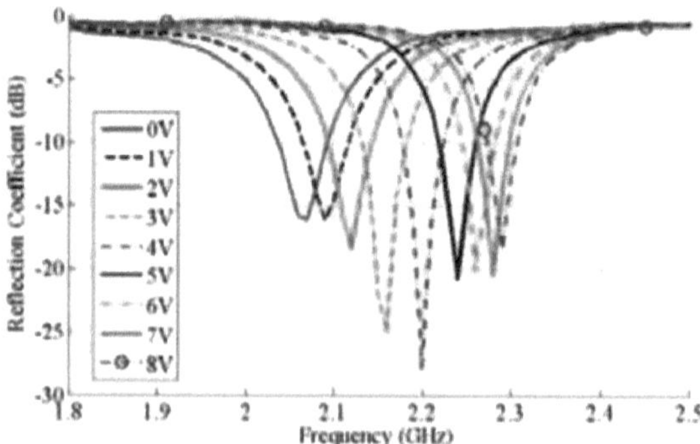

Figure III.42: Measured S11 parameters of the proposed antenna with variable DC voltage across the varactor diode.

When no voltage has ёlё applied across the varactor diode:
- Its reactive impedance is 10 pF and ;
- The antenna had a resonance frequency of 2.06 GHz and ;
- Its bandwidth is 50 MHz.

When DC voltage has been applied across the varactor diode:
- Its reactive impedance a decreases and falls below 1 pF (at 7V) and ;
- The resonance frequency is shifted to the higher side and ;
- Its minimum bandwidth remains 50 MHz.

Increasing the voltage across the terminals of the varactor diode causes a gradual change in the antenna's resonant frequency. A shift of 40 MHz, on average, in the resonance frequency has been observed.

Frequency reconfigurability and miniaturisation was achieved by using a varactor diode in the antenna ground plane. The antenna had a bandwidth of 50 MHz and by applying external DC voltage, it could switch between several bands covering 2 GHz to 2.3 GHz to be useful for several applications, including cognitive radios.

Another study has ёlё proposёe for cognitive radio front-end systems in 1 paper [41]. The proposed antenna has a miniaturised size of $25*30x0.762$ mm^3 and uses four varactor diodes to control the resonant frequency of the antenna (Figure III.43).

The resonant frequency of the antenna can be adjusted electronically by changing the effective electrical length of the resonant slot, which is achieved by employing varactor diodes within the slot. In addition, the simple polarisation of the varactor diodes has little effect on the performance of the antenna.

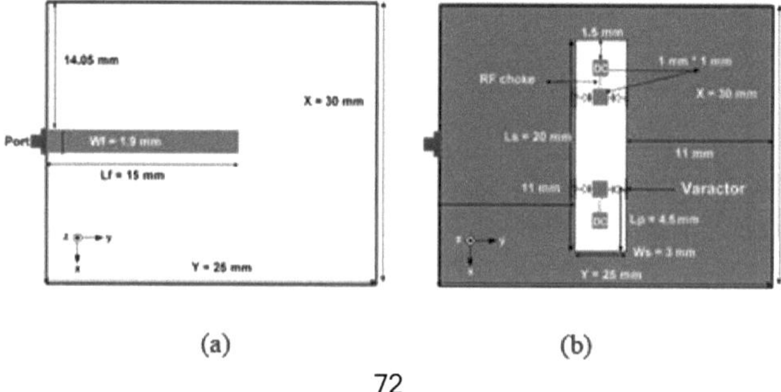

(a)   (b)

Figure III.43: Geometry of the proposed antenna (a) Top surface, (b) Bottom surface
The simulation results (see Figure III.44) show that the proposed antenna is capable of changing frequency over a 0.93 GHz adjustment range (from 1.83 to 2.76 GHz) with a frequency adjustment range of 40.5%.

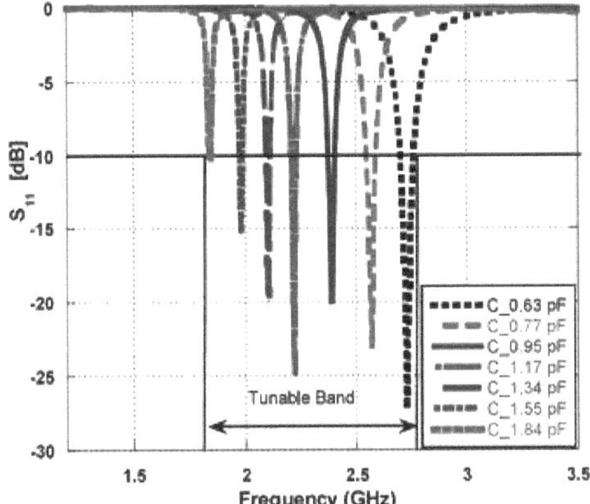

Figure III.44: Simulated S11 parameters of the proposed antenna.

c) Inductive load

In paper [42], it is shown that the size of a planar Inverted F Antenna (IFA) for integration into the scan of a mobile device is reduced by inductive or capacitive loading.

The inductively loaded IFA is constructed as shown in Figure III.45(a). The geometrical parameters are fixed and different inductors are placed on the IFA.

The resonance frequency without inductive load is 5.2 GHz (X = 5.77 cm). The bandwidth and resonance frequency of the unloaded antenna are suitable for WLAN.

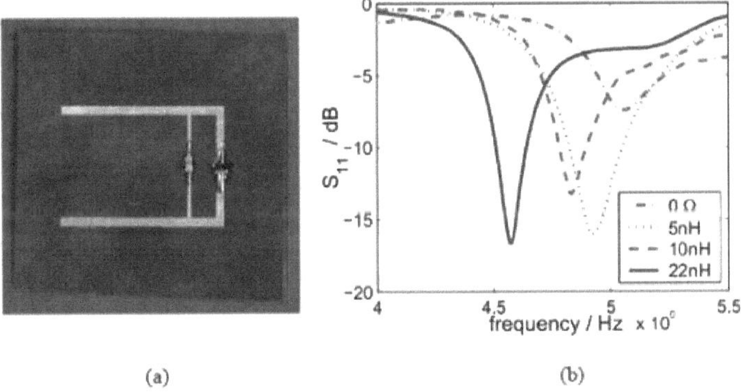

(a)   (b)

Figure III.45: Photograph of the used IFA loaded with an inductor. (b) Rësults of reflection coefficient measurements for different inductive loads.

Using an inductive load with 5nH, 10nH and 22nH, the resonance frequency can be reduced to 650 MHz. The variation in matching (see Figure III.45-b) is a function of the matching network for all measurements. The matching can be improved by a specific matching network for each antenna.

To build an antenna that operates at 5.2 GHz, the size of the antenna must be reduced. The gain measured in the direction of the main beam is always approximately the same with and without a 22 nH inductive load, about 2 dBi.

6.4. Technique 4: Short circuit

6.4.1. Description of the technique

The addition of short-circuit structures, by introducing a new current path, modifies the field distribution of the antenna, resulting in the generation of a new resonance frequency in the low-frequency band.

The use of short-circuit terminals between the radiating ferment and the ground plane and the incorporation of a combination of different techniques can improve the performance of the microstrip antenna in terms of bandwidth or polarisation purity.

6.4.2. Work carried out and discussion

A new dual-band printed antenna for WLAN application covering the 2.4 GHz and 5.8 GHz bands has been presented in reference [43]. Short-circuit technology is used to miniaturise the antenna. The L-shaped slot is used to generate dual bands.

Figure III.46 illustrates the geometry of the dual-band patch antenna.

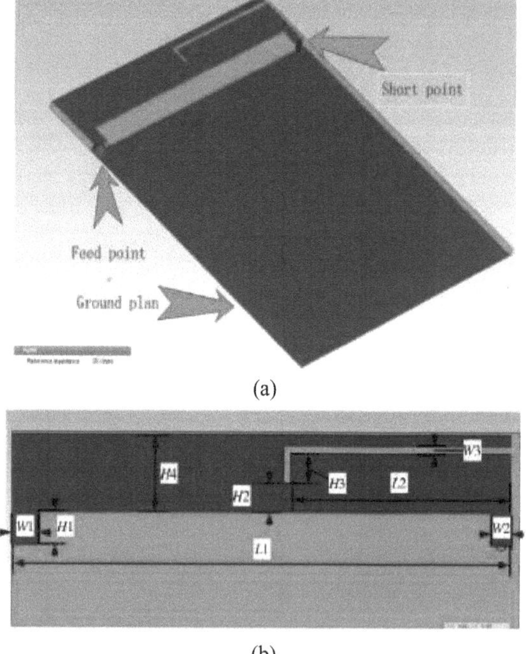

(a)

(b)

Figure III.46: Gëomëtrie of the proposed antenna. (a) Free view of the antenna. (b) Top view of the antenna.

The antenna is printedëe on a FR4 PCB substrate board with a thickness of 1 mm and a relative permittivity of 4.5. The antenna feed point is representedësentë on the left and the short-circuit point on the right. Dual band operation can be generated using an L-shaped slot.

Reference [44] presents a simple monopole slot printed antenna with two parasitic short-circuited bands (long and short) for a wide operating area of a penta-band wireless network on a slim mobile phone.

The monopole slot is 40 mm long and is positioned at a distance of 6.5 mm from the ground plane of the mobile phone (Figure III.47).

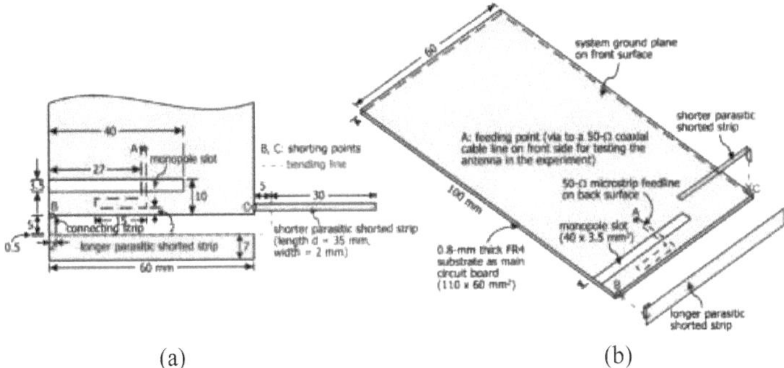

(a)   (b)

Figure III.47: (a) Dimensions of the antenna in its planar structure, (b) Gëomëtrie of the proposed monopole slot antenna.

To analyse the operating principle of the antenna, Figure III.48 shows the measured reflection coefficient of the proposëe antenna, the case with a monopole slot only (notë Ref.1) and the case with a monopole slot and a long short-circuited parasitic band only (Ref. 2).

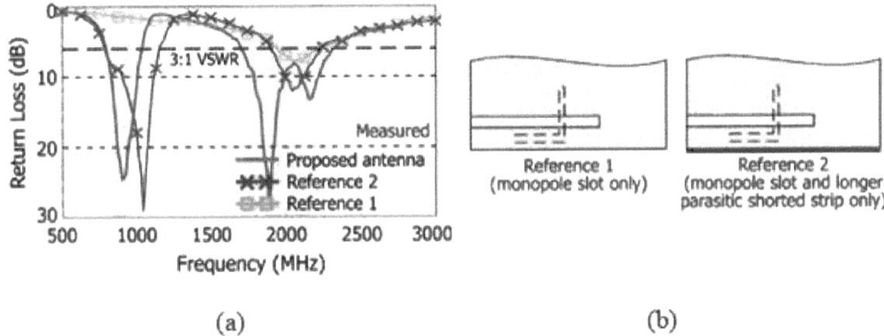

(a)   (b)

Figure III.48: (a) Reflectance coefficient measured for the proposed antenna. (b) Ref.1: monopole slot only or; Ref.2: monopole slot with
long short-circuited stray band only.

For Ref. 1, a quarter-wave resonance mode at about 2 GHz is only excited. When the long-short parasitic band is added to form Ref. 2, a chassis dipole resonance mode has occurred at about 1 GHz, which covers the desired 824-960 MHz band. Finally, by adding a short parasitic band to form the proposed antenna, an additional quarter-wave mode at about 2.1 GHz is generated. As a result, two higher order resonant modes at about 1.9 and 2.1 GHz are generated to combine with the resonant modes provided by refs. 1 and 2 to form a broadband antenna top band to cover the desired 1710-2170 MHz band.

The addition of the two short-circuited parasitic bands (short and long) connected to the bottom corner of the ground plane, allows the monopole slot antenna to be miniaturised (bringing additional resonant modes to the antenna) and to achieve good excitation. The excited resonance modes form two wide operating bands to cover, respectively, the GSM850/900 (824-960 MHz) and GSM1800/1900/WCDMA (17102170 MHz) bands.

A short-circuited rectangular microstrip antenna (RMA) with a defective patch surface (oblique slot) is proposed and experimentally studied in [45] for extended bandwidth and improved cross-polarised (XP) radiation compared with the principal polarised (CP) radiation pattern without changing the latter.

Isolation of around 23-35 dB between CP and XP radiation with a 25% bandwidth is achieved with the proposed structure. The measured gain of the antenna is stable at around 6.2 dBi over the entire band.

The RMA antenna is designed using a thin strip of copper with a thickness of 0.1 mm, a length of L=7 mm and W=12.12 mm to operate around 13.4 GHz. PTFE material ($_{sr=52.}$33) with a thickness of 5.58 mm is used as the substrate. The size of the ground plane is 80*80 mm .^{2}

A pair of wide slits cut into the surface of the patch gives rise to a defective patch structure, as shown in Figure III.49-b.

Four grooves are cut in the corners of the patch and four metal shorting pins are allowed to pass through these grooves, welded to the patch corners and thus shortening the patch corners to the ground plane.

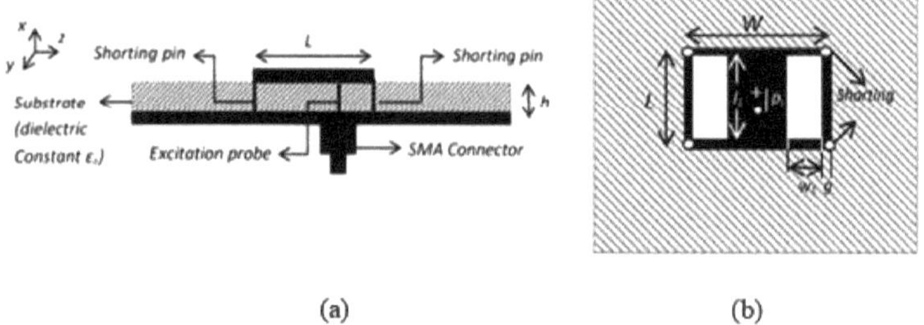

(a)   (b)

Figure III.49: Schematic diagram of the RMA structure: (a) side view, (b) top view.

A clear physical indication of the observed improvement in XP radiation can also be seen in the distribution of the electric field over the surface of the patch, which is shown in Figure III.50.

The literature shows that the main sources of XP are radiation from the non-radiating edges and patch corners of an RMA. Therefore, to reduce the radiation from the non-radiating edges, the field amplitude must be minimised. Figure III.50 clearly shows that the amplitude of the electric field is significantly lower at the non-radiating edges as well as at the patch corners in the case of the final structure compared with the conventional RMA.

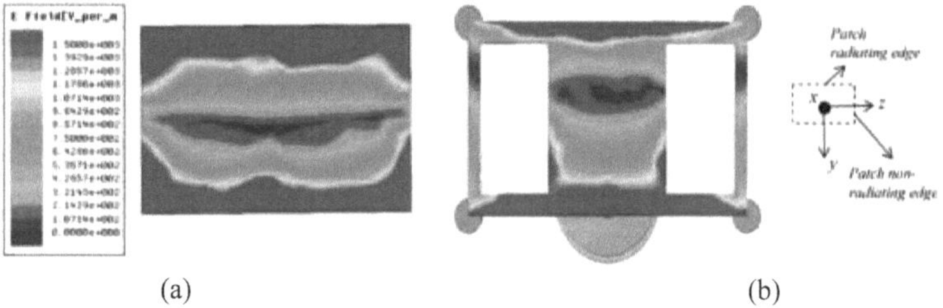

(a)   (b)

Figure III.50: Variations in the amplitude of the electric field over the patch surface for:
(a) conventional RMA, (b) proposée

structure.

The field strengths at the radiating edges are similar in both cases and therefore have no influence on the main radiation pattern (co-polarisation).

Figure III. 51 shows the reflection coefficients measured for the two structures (conventional and final).

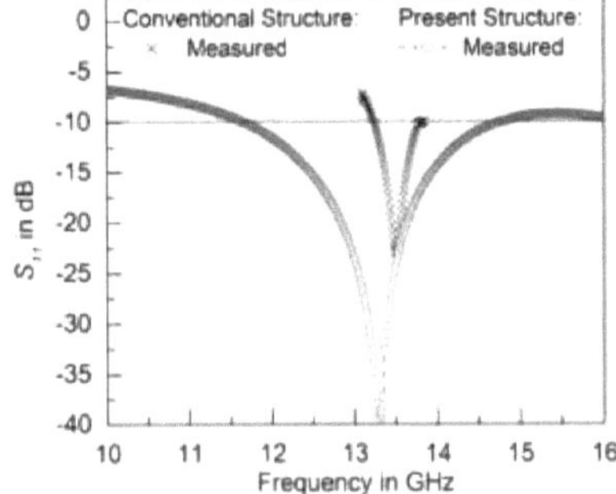

Figure Ш.51: Reflection coefficient for the conventional structure and the final structure
(short-circuited RMA loaded with a wide-slit dipole with defective patch surface)

The introduction of large slots on the patch surface improves the bandwidth to -10 dB of the final antenna structure compared to the conventional one. The result measures rëyëк that the present antenna produces a wide bandwidth of 25% (from 11.50 GHz to 14.84 GHz) whereas that for a conventional antenna is only 5%.

The simulation results for bandwidth, maximum CP gain and CP- XP isolation for the proposed antenna with a varied slot width (w1) are shown in Table III.3.

Table Ш.3: Bandwidth (-10 dB) and CP-XP isolation for the proposée antenna for different widths (w1) from slot to patch surface.

Type of structure	w1 (mm)	bandwidth (%)		Maximum CP gain at resonance (dB)		minimum insulation CP- XP (dB)	
		Simulee	measured	Simulate	measure	Simulee	measured
Conventional RMA	No slots	5	6	6.2	6.07	11	10.05
RMA structure with defective patch surface and no short circuits	3.2	25.38	24.6	6.2	6.15	19.18	19.01
RMA structure with a faulty slotted patch surface and short-circuit pins	3.2	25.01	24.89	6.18	6.13	23.2	22.5

According to Table Ш.3, the bandwidth of the RMA structure with the defected patch surface becomes almost 25%, while the same for a conventional patch is only 5%. The RMA with defective patch surface having a slot width w1=3.2 mm also shows a significant improvement in the isolation of the CP-XP and it is about 19 dB while the same for the conventional patch is only 10.05 dB. Now, for a further improvement in polarisation purity, four shorting pins are used at the four corners of the patch with the same slot width w1.

7. Conclusion

In conclusion, it can be said that the use of these four techniques, such as the modification of the ground plane (DGS) with loading by very high permittivity materials (BST) and the

77

integration of localised ёlёments, as well as the short-circuiting technique, or the combination between these techniques makes it possible to achieve a significant miniaturisation of planar antennas, to improve the purity of polarisation and to obtain a wide frequency band.

Bibliography of Chapter III

[1]   Riki H Patel, Arpan Desai, Trushit Upadhyaya, "*a discussion on electrically small antenna property*", Microwave and Optical Technology Letters, Vol. 57, No. 10, Pp. 2386-2388, Oct. 2015.

[2]   S.A. Schelkunoff, H.T Friis, '*Antennas and theory*', Chapter 10, Wiley, New York, NY, 1952.

[3]   L. Wang, M.Q. Yuan, Q.-H. Liu, "*A dual-band printed electrically small antenna covered by two capacitive split-ring resonators*", IEEE Antennas Wireless Propag Lett 10 (2011), 824-826.

[4]   John L. Volakis, Chi-Chili Chen, K yohei Fujimoto, "*Small Antennas: Miniaturization Techniques & Applications*", McGraw-Hill Companies, USA, 2010.

[5]   R. C. Hansen, "*Fundamental limitations in antennas*," Proceedings of the IEEE, vol. 69, no. 2, February 1981, pp. 170-182.

[6]   G. A. Thiele, P. L. Detweiler, and R. P. Penno, "*On the lower bound of the radiation Q for electrically small antennas*," IEEE Transactions on Antennas and Propagation, vol. AP-51, June 2003, pp. 1263-1269.

[7]   R. F. Harrington, "*Effect of antenna size on gain, bandwidth, and efficiency*", Journal of Research of the National Bureau of Standards, vol. 64D, January-February 1960, pp. 1-12.

[8]   R. W. P. King, *The Theory of Linear Antennas*, Harvard University Press, Cambridge, Mass. 1956.

[9]   S. R. Best, "*The performance properties of electrically small resonant multiple-arm folded wire antennas*," IEEE Antennas and Propagation Magazine, vol. 47, no. 4, August 2005, pp. 1327.

[10]   H.A. Wheeler, Fundamental limitations of small antennas, Proc IRE 35 (1947), 1479-1484.

[11]   R. W. P. King, *The theory of liner antennas*, Harvard University Press, Cambridge, MA, pp. 184-192, 1956.

[12]   Muhammad Umar Khan, Mohammad Said Sharawi, Raj Mittra, "*Microstrip patch antenna miniaturisation techniques: a review*", IET Microwaves, Antennas & Propagation, Vol. 9, No. 9, pp. 913-922, 2015.

[13]   S.M.A.M.H. Abadi and N. Behdad, *An electrically small, vertically polarized ultrawideband antenna with monopole-like radiation characteristics*, IEEE Antennas Wireless Propag Lett 13 (2014), 742-745.

[14]   M.T. Ali, N. Nordin, I. Pasya, and M.N. Md Tan, *H-shaped micro-strip patch antenna using L-probe fed for wideband applications*, 6th European Conference on Antennas and Propagation (EUCAP), pp. 2827-2831, Prague, 2012.

[15]   L. H. Weng, Y. C. Guo, X. W. Sh i, and X. Q. Chen, "*An overview on defected ground structure*," Progress In Electromagnetics Research B, Vol. 7, 173-189, 2008.

[16]   C. Insik, and L. Bomson, "*Design of defected ground structures for harmonic control of active microstrip antenna*," IEEE Antennas Propag. Soc. Int. Symp. vol. 2, 852-855, 2002.

[17]   J. S. Park, J. H. Kim, J. H. Lee, et al, "*A novel equivalent circuit and modeling method for defected ground structure and its application to optimization of a DGS low-pass filter*," IEEE Microwave Symposium Digest, 2002 IEEE MTT-S International, vol. 1, 417-420, 2002.

[18]   G.-L. Wu, W. M., X.-W. Dai, et al, "*Design of novel dual-band bandpass filter with microstrip meander-loop resonator and DGS*," Progress in Electromagnetics Research , PIER 78, 17-24, 2008.

[19]   Amiya B. Sahoo, Ayush Biswal, Chandan K. Sahu, Jogesh C. Dash, B. B. Mangaraj, "*Design of multi-band rectangular patch antennas using defected ground structure (DGS)*", 2nd IEEE International Conference on Recent Trends in Electronics, Information & Communication Technology (RTEICT), Bangalore, India, 19-20 May 2017.

[20]   R. Er-rebyiy, J. Zbitou, A. Tajmouati, M. Latrach, A. Errkik, L. El Abdellaoui, "*A new design of a miniature microstrip patch antenna using Defected Ground Structure DGS*", International Conference on Wireless Technologies, Embedded and Intelligent Systems (WITS), Fez, Morocco, 19-20 April 2017.

[21]   Vrishali Mahesh Belekar, Prachi Mukherji, Mahesh Pote, "*Improved microstrip patch antenna*

with enhanced bandwidth, efficiency and reduced return loss using DGS", International Conference on Wireless Communications, Signal Processing and Networking (WiSPNET), Chennai, India, 22-24 March 2017.

[22]   K. Wei, J. Y. Li, L. Wang, R. Xu, Z. J. Xing, "*A New Technique to Design Circularly Polarized Microstrip Antenna by Fractal Defected Ground Structure*," IEEE Transactions on Antennas and Propagation, Vol. 65, No. 7, Pp. 3721 - 3725, July 2017.

[23]   Korany R. Mahmoud, Ahmed M. Montaser, "*Optimised 4^4 millimetre-wave antenna array with DGS using hybrid ECFO-NM algorithm for 5G mobile networks*", IET Microwaves, Antennas & Propagation, Vol. 11, No. 11, Pp. 1516 - 1523, 9 Aug. 2017.

[24]   Li Gu, Yan-Wen Zhao, Qiang-Ming Cai, Zhi-Peng Zhang, Bi-Hui Xu, Zai-Ping Nie, "*Scanning Enhanced Low-Profile Broadband Phased Array With Radiator-Sharing Approach and Defected Ground Structures*", IEEE Transactions on Antennas and Propagation, Vol. 65, No. 11, Pp. 5846 - 5854, Nov. 2017.

[25]   Chandrakanta Kumar, Mahammad Intiyas Pasha, Debatosh Guha, "*Defected Ground Structure Integrated Microstrip Array Antenna for Improved Radiation Properties*", IEEE Antennas and Wireless Propagation Letters, Vol. 16, Pp. 310 - 312, 30 May 2016.

[26]   Suleyman Kuzu, Nursel Akcam, "*Array Antenna Using Defected Ground Structure Shaped With Fractal Form Generated by Apollonius Circle*", IEEE Antennas and Wireless Propagation Letters (Volume: 16, Pp. 1020 - 1023, 12 Oct. 2016.

[27]   Kedar Trivedi, Dhaval Pujara, "*Mutual coupling reduction in wideband tree shaped fractal dielectric resonator antenna array using defected ground structure for MIMO applications*", Received: 12 April 2017, DOI: 10.1002/mop.30810

[28]   M.C. Lim, S.K.A. Rahim, M.R. Hamid, P.J. Soh, Aa Eteng, "*Semi-transparent frequency reconfigurable antenna with DGS*", Received: 4 June 2017, DOI: 10.1002/mop.30915

[29]   Yatendra Kumar, Ravi Kumar Gangwar, Binod Kumar Kanaujia, "*Compact broadband circularly polarized Hook- shaped microstrip antenna with DGS plane*", Received: 27 November 2017, Revised: 14 February 2018, Accepted: 14 February 2018, DOI: 10.1002/mmce.21275

[30]   Sumy Mathew, Mailadil T. Sebasian, Pezholil Mohanan, "*A low profile, high permittivity cylindrical dielectric resonator antenna for microwave communication*", International Conference on Advances in Computing and Communications, Pp. 267-269, 2012.

[31]   Antrisha Daneraichi Setiawan, Achmad Munir, "*Incorporation of High Permittivity Circular Dielectric Resonator for Enhancing Resonant Frequency of Microstrip Antenna*", 15th Intl. Conf. QiR: Intl. Symp. Elec. and Com. Eng, Pp. 87-90, 2017.

[32]   Ying Liu, Hu Liu, Ming Wei, Shuxi Gong, "*A Low-Profile and High-Permittivity Dielectric Resonator Antenna With Enhanced Bandwidth*", IEEE Antennas and Wireless Propagation Letters, VOL. 14, Pp. 791-794, 2015.

[33]   Idris Messaoudene, Farouk Chetouah, Massinissa Belazzoug, "*Compact Rectangular DRA with High Permittivity Stacked Resonator for RADAR Applications*", 7th SEMINAR On Detection Systems: Architectures And Technologies (DAT'2017), February 20-22, 2017, Algiers, Algeria.

[34]   Wee Fwen Hoon, Mohd Fareq bin Abd. Malek, Liew Hui Fang, Lee Yeng Seng, liyana Zahid, "*The Miniaturization of High Permittivity DRA with Array Patches*", 2013 First International Conference on Artificial Intelligence, Modelling & Simulation, Pp. 443-445, 2013.

[35]   Y. Hwang, Y. P. Zhang, T. K. Lo, "*Planar inverted-F antennas loaded with very high permittivity ceramics*", Radio Science, VOL. 39, RS2002, Pp. 1-10, 2004.

[36]   J. L. G. Medeiros, A. G. d'Assun^ao, L. M. Mendon^a, "*Microstrip Fractal Patch Antennas Using High Permittivity Ceramic Substrate*",

[37]   Yu-suke Takigawa, Shinya Kashihara, Futoshi Kuroki, "*Integrated Slot Spiral Antenna Etched on Heavily-High Permittivity Piece*", Proceedings of Asia-Pacific Microwave Conference, 2007.

[38]   Braham Chaouche Y, Bouttout F, Messaoudene I, Pichon L, Belazzoug M, Chetouah F.

,'*Design of reconfigurable fractal antenna using pin diode switch for wireless applications*, 16th Mediterranean Microwave Symposium (MMS'16), 14-16 November2016.

[39]   BorhaniRezaei MP, Valizade A. *Design of a reconfigurable min-iaturized microstrip Antenna for switchable multiband systems*', IEEE Antennas Wireless Propag Lett. 015;15:822- 825.

[40]   Muhammad U. Khan, Rifaqat Hussain, Mohammad S. Sharawi, "*A Compact Reconfigurable and Miniaturized Patch Antenna*", Pp. 140-141, 4th Asia-Pacific Conference on Antennas and Propagation (APCAP), 2015.

[41]   Hany A. Atallah, Adel B. Abdel-Rahman, Kuniaki Yoshitomi, Ramesh K. Pokharel, "*Design of miniaturized reconfigurable slot antenna using varactor diodes for cognitive radio systems*", 4th International Japan-Egypt Conference on Electronics, Communications and Computers (JEC-ECC), Pp. 63-66, Cairo, Egypt, 31 May-2 June 2016.

[42]   S. Schulteis, C. Waldschmidt, W. Sorgel, W. Wiesbeck, "*A Small Planar Inverted F Antenna with Capacitive and Inductive Loading*", IEEE Antennas and Propagation Society International Symposium, Monterey, CA, USA, 20-25 June 2004.

[43]   Liu H, Zhang L, Pan J, Liu C, Lin Z., "*A novel dual-band antenna for WLAN application*", Progress in Electromagnetic Research Symposium (PIERS), Pp. 484-486, China. 8 -11 Aug. 2016.

[44]   Chih-Hua Chang, Wan-Chu Wei, Pei-Ji Ma, Shao-Yu Huang, "*Simple printed WWAN monopole slot antenna with parasitic shorted strips for slim mobile phone application*", Microwave and Optical Technology Letters, Vol. 55, No. 12, Pp. 2835-3541, December 2013.

[45]   Abhijyoti Ghosh, Sudipta Chattopadhyay, L. Lolit Kumar Singh, Banani Basu, "*Wide bandwidth microstrip antenna with defected patch surface for low cross polarization applications*", Int J RF Microw Comput Aided Eng, Pp. 1-10, 2017, https://doi.org/10.1002/mmce.21127

Miniaturisation of a rectangular rectangular dielectric resonator by loading a thin film of BST material with very high permittivity

1. Introduction

The performance of printed dées antennas depends very much on the substrate on which they are mounted. The substrate not only provides mechanical support for the antenna but also affects properties such as resonant frequency, bandwidth and, most importantly, radiation efficiency, the thickness and permittivity of the substrate influence the bandwidth of the antenna. A suitable substrate must therefore meet the mechanical and electrical requirements of the design.

Parameters such as high dielectric constant, low dielectric loss, high charge storage capacity and low leakage current density are important factors in the choice of dielectric material.

Barium strontium titanate (BaxSr1-xTiO3 or BST) [1,2] is one of the most popular materials for microwave applications, due to its advantages such as chemical stability, good temperature behaviour, high electric field dependent dielectric permittivity, high tunability and low loss tangent at microwave frequencies [3].

Ferroelectric materials have been applied to many microwave component designs, such as tunable antennas and phased arrays based on BST varactors, phase shifters and tunable filters. In particular, BST thin film is considered one of the most promising dielectric materials for high-density capacitors in DRAMs with high charge storage density.

This chapter is devoted to the study and analysis of a miniaturised rectangular dielectric resonator antenna loaded with a thin film of very high permittivity BST ferroelectric material.

BST ferroelectric materials are briefly presented in section 2 of this chapter. In Section 3, a dielectric resonator antenna based on a rectangular structure is presented. Simulation results are discussed and compared before and after the integration of a thin layer of BST material in order to achieve significant miniaturisation. The conclusion is given in section 4.

The work presented in section 3 of this chapter was the subject of an international paper [4] which was published and presented at the conference in Al-Ain, United Arab Emirates.

2. Dielectric properties used in the manufacture of RD antennas

2.1. Properties of dielectric substrates

2.1.1. Dielectric materials [5,6]

The critical element that has enabled the development of microstrip patch antennas is the innovation of substrate materials. This specific material can influence the properties of the antenna.

ëelectrics of the antenna, the transmission line as well as the circuits. Thus, a appropnë substrate must satisfy both mëcanical and ëelectrical requirements at the same time.

A diëlectric matëriau is known as a poor conductor of ëlectricitë, but a good means of ëlectrostatic current flow. When a ëlectromagnëtic wave moves through a diëlectric, the speed of the wave will be reduced and it will behave as if it had a shorter wavelength.

Materials can be classified according to their permittee. Those with a positive real part diëlectrical constant are diëlectrics. Metals are good electrical conductors and have an egative permittance at very high frequencies, in which there is no propagation of electromagnetic waves.

The permittee of the diëlectrical is in gënëral complex and is donime by:

$$\varepsilon = \varepsilon' - j\varepsilon''$$
(4.1)

Where; ε' is the real part which represents the relative permittivity of the matëriau (diëlectrical constant) and ε'' is the imaginary part and indicates the loss factor woven into it such that:

$$\varepsilon'' = \frac{\sigma}{\varepsilon_0 \omega} d$$
(4.2)

In this expression, σ is the total conductivity of the material, ε_0 is the permittee of free space and ω is the angular frequency of the field.

I.'one conductive in the SI is Siemens per metre (S/m) which assumes that in the above expression so is expressed in farads per metre (F/m) and (') in radians per second.

The dielectric properties are determined as values ε' et ε'' or relative permittivity and loss tangent which is relevant to both ε' et ε'' as a function of frequency.

2.1.2. Permit

The permittee of a medium can be consideredërëe as the quale of a matëriau that allows it to store ëlectric charge. The units of permittivity are Farads / metre (F/m). The permittivity in vacuum (free space) is noted so, its value is 8.854×10^{12} F/m. Materials other than vacuum have a permittivity greater than so, and are often dësignës by their relative permittivity sp

$$\varepsilon_r = \frac{\varepsilon}{\varepsilon_0}$$
(4.3)

Or; ε: middle permittee, Fm^{-1}

ε_r: relative permittee of the environment (dimensionless)

In microwaves, the relative permittivity £r is often referred to as the "diëlectric constant". The diëlectrical constant is a function of frequency and it is important to characterise the substrate matëriau for the range of operating frequencies.

Diëlectric materials with minimal losses are prëfërësed to have maximum antenna radiation efficiency. However, there are always losses associated with diëlectrics.

For two flat plates filled with a dielectric material, the permittivity of the material can be written in:

$$\varepsilon = \varepsilon_r \varepsilon_0 = \frac{Q}{EA}$$
(4.4)

Q: the load that is uniformly distributed between the plates, (C)

E: electric field, $(Vm)^{-1}$

A: surface area of the plate, $(m)^2$

As a result, a given quantity of material with a high permittivity can store more charges than a material with a lower permittivity. A high permittivity tends to reduce any electric field present. The capacitance C between the plates is given by:

$$C = \frac{\varepsilon A}{d} = \frac{\varepsilon_r \varepsilon_0 A}{d}$$ (4.5)

Or ; d: distance between the two plates, (m)

In this way, the capacitance of a capacitor can be increased by increasing the permittivity of the dielectric material inside.

To quantify the losses of dielectric materials, we use another term known as the loss tangent, which is explained in the next section.

2.1.3. Loss tangent

The loss tangent, tanS (also known as the dissipation factor) defines the loss of the medium and characterises the amount of power transformed into heat in the material. It is given by the negative ratio of the imaginary part to the real part of the material's permittivity at any particular frequency. The loss tangent is given by the relationship:

$$tan\left(\delta(\omega)\right) = \frac{\varepsilon''(\omega)}{\varepsilon'(\omega)}$$ (4.6)

A large loss tangent means that the material has a lot of dielectric absorption and a high loss of power transmitted through the dielectric.

Since the radiation efficiency of an antenna is strongly Hëc to the loss factor of the matëriau, a large loss tangent means a very low radiation efficac^ of the antenna.

The relative permittiv^s of some common matërials are ënumërëes in Table IV. 1. Note that they are functions of frequency and tempërature. Normally, the higher the frequency, the smaller the permittivity^ in the radio frequency band. It should also be pointed out that almost all conductors have a relative permittivity of 1.

Table IV.1: Relative permittivity of some common materials at 100 MHz. [7]

Material	Relative licence	Material	Relative licence
ABS (plastic)	2.4-3.8	Polypropylene	2.2
Air	1	Polyvinyl chloride (PVC)	3
Alumina	9.8	Porcelain	5.1-5.9
Aluminium silicate	5.3-5.5	PTFE-teflon	2.1
Balsa wood	1.37 to 1 MHz 1.22 to 3 GHz	PTFE-ceramic	10.2
Concrete	~8	PTFE-glass	2.1-2.55
Copper	1	RT / Duroid 5870	2,33
Diamond	5.5-10	RT / Duroid 6006	6.15 a 3GHz
Epoxy (FR4)	4,4	Rubber	3.0-4.0
PCB epoxy glass	5.2	Sapphire	9.4
Ethyl alcohol (absolute)	24.5 to 1 MHz 6.5 to 3 GHz	Sea water	80
FR-4 (G-10) -low resin -high resin	4.9 4.2	Silicon	11.7-12.9
GaAs	13.0	Soil	~10
Glass	~4	Soil (dry sandy)	2.59 at 1MHz 2.55 to 3 GHz
Gold	1	Water (32T) (68T) (212T)	88,0 80,4 55,3
Ice (pure distilled water)	4.15 a 1 MHz	Wood	~ 2

	3.2 to 3 GHz		

2.2. Properties of resonators with a very high dielectric permittivity [8,9].

2.2.1. Ferroelectric materials

Fen^electric materials belong to a class of perovskite oxide materials that exhibit a spontaneous ëlectric polarisation below the Curie temperature T_c.

On cooling, there is a transition from the pa^electric ëtate to the fen^electric ëtate at T_c, where the maximum dïelectric constant is observed. At temperatures below Tc (i.e. at the fen^electric ëtate), spontaneous ëlectric polarisation is present when there is relative displacement of ions in the fen^electric materials; this results in a net dipole moment. The orientation of the dipole moment in a fen^lectrical material can be In this case, the polarisation of the material changes from one orientation to the other under the influence of an applied ëelectric field, causing a change in the dielectric constant of the material [10-13]. The relationship between polarisation and applied electric field is represented by the hysteresis curve in Figure IV.1:

- In the ferroelectric state, the polarisation of the ferroelectrics remains even in the absence of an applied electric field.

- In the paraelectric state, there is only one polarisation with the application of an external electric field.

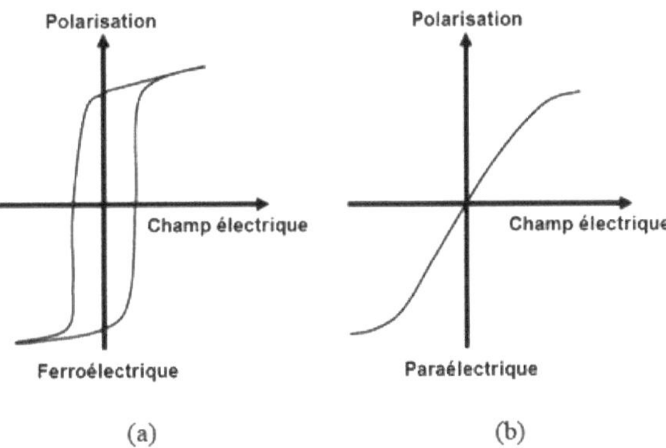

Figure IV.1: Polarisation of a ferroelectric material in the ferroelectric and paraelectric states in response to an applied external electric field, showing: (a) hysteresis, and (b) no hysteresis.

Examples of ferroelectric materials are barium titanate ($BaTiO_3$) and barium strontium titanate (BaxSr1-xTiO3 or BST). Figure IV.2 shows the perovskite structure and the electrical polarisation of a BST unit cell (i.e. the displacement of the Ti atom) in response to an external electric field.

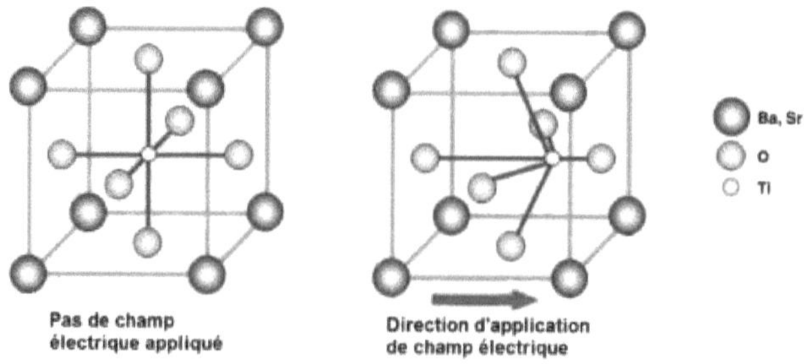

Figure IV.2: Perovskite structure and electrical polarisation of a unit cell of barium-strontium (Ba, Sr) TiO3 Titanate in response to an applied external electric field . [14]

An operating tempërature slightly above the Curie tempërature in the paraelectric phase is normally preferred for tunable microwave devices because ferroelectrics in this state are free from the hysteretic effect and have moderate loss [11, 12]. Despite this technical distinction, ferroelectrics used in the paraelectric phase for tunable microwave devices are still referred to as ferroelectric materials.

2.2.2. Dielectric properties

Ferroelectrics are essentially non-linear dielectrics because their dielectric constant depends on the external electric field and temperature.

a. Field-dependent dielectric constant

Several ferroelectric materials are suitable for integration with microwave monolithic integrated circuits (MMICs) and the most widely studied material for microwave application is BaxSri-xTiO3 (BST), where x can vary from 0 to 1.

BST is a composition in which barium ions are introduced into pure strontium titanate (SrTiO3 or STO) in an attempt to improve microwave properties. Pure STO is particularly attractive because of its crystalline compatibility with high-temperature superconductors (HTS) and it is in the paraelectric state at any temperature, so there is no Curie temperature above 0 K. On the other hand, barium titanate has a Curie temperature of around 400 K.

A different composition ratio allows the Curie temperature to be adapted. Typically, a value of x = 0.4 - 0.6 is used to optimise its properties at room temperature, and a value of around x = 0.1 - 0.2 is used when the material is used in conjunction with HTS films [10,15, 16-18].

Figure IV. 3 shows the variation of the relative dïelectric constant Sr of Ba0.5Sr0.5TiO3 (BST) thin film as a function of electric field and temperature [19].

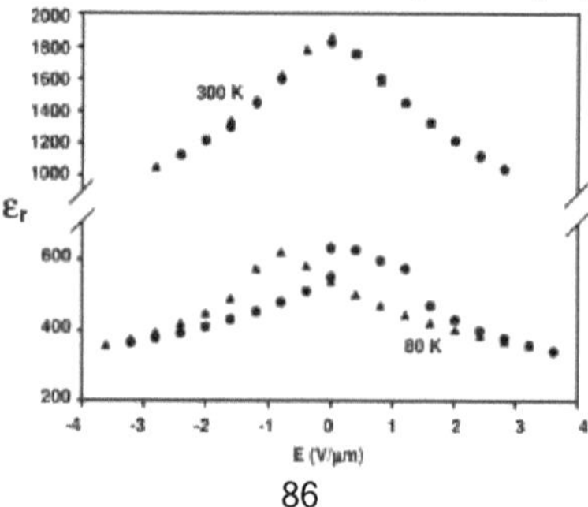

Figure IV.3: Dielectric constant of a thin film of BST at 80 K and 300 K [19].

As shown in Figure IV. 3, the dïelectrical constant of BST decreases with increasing ëlectric field and for tempërature:

- At 300 K, BST thin films have a ;:,■ in the range 1000-2000 without hystërësis [19]. This shows that the thin films are in the paraëelectric state.

- As the tempërature drops to 80 K, Sr drops below 700 and hystërësis is observed as the voltage is sweptë up and down. This is expected as the BST thin film is in the ferroelectric state.

b. Tunability

One of the most important properties of ferroelectric material is the strong dependence of its dielectric constant on the applied external electric field E. This characteristic is generally described by the tunability n, defined as the ratio of the dielectric constant of the material at zero electric field to its dielectric constant at a non-zero electric field, as expressed by equation (4.7). The relative tunability n in percent is defined by equation (4.8).

$$n = \frac{\varepsilon(0)}{\varepsilon(E)} \tag{4.7}$$

$$n(\%) = \frac{\varepsilon(0) - \varepsilon(E)}{\varepsilon(0)} \cdot 100 \tag{4.8}$$

c. Loss tangent

Ferroelectric materials, like all other dielectrics, generally suffer from dielectric losses. Relation (4.1) expresses mathematically, the dielectric constant of ferroelectrics which can be represented in complex form with both real s' and imaginary s" parts. The imaginary part of the dielectric constant s" takes into account the loss factor due to the polarisation delay when the electric field is applied and the energy dissipation associated with charge polarisation. This loss factor is characterised by the ratio between the imaginary part and the real part of the dielectric constant, commonly referred to as the loss tangent, tan 6 (relation (4.6)).

Materials with high tunability and low loss tangent at microwave frequencies are highly desirable in microwave engineering applications. However, in conjunction with high tunability, most ferroelectrics also possess high loss and temperature dependence [20, 21]. The correlation between tunability and loss tangent often forces designers to choose the material with the optimum compromise between these two parameters for best device performance.

2.23. **Bulk, thick film and thin film**

The ieTroelectric materials come in several forms, namely bulk, thick film and thin film. Each of these forms has its advantages and inconvënients:

a. Bulk: Bulk ieiToeiectrics are typically 500 to 2000 pm thick. Due to their very ëlevëe dïelectric constant, typically of the order of tens to thousands, bulk ferroelectrics are useful in substantially reducing the size of microwave devices.

Bulk ferroelectrics have been used in many applications such as tunable dïelectric resonators [22, 23], tunable filters [24, 25], varactors [26] and lens antennas [27].

However, 1 major drawback of bulk ieTroelectrics is that very high tuning voltages, of the order of hundreds of volts to tens of kilovolts [28, 12], are required to obtain a usable tuning range. This greatly limits their use in tunable microwave devices.

Nevertheless, bulk ferroelectrics generally exhibit lower loss tangent values than other forms of ferroelectrics.

b. Thick films: Thick films refer to a thickness greater than about 1 pm. In addition, they are generally polycrystalline. Thick ferroelectric films are considered to be a more practical form than bulk because a significantly lower tuning voltage is required (a few hundred volts). Furthermore, with the development of tape casting [29] or screen printing technology [30], the production cost of thick films has decreased considerably, making thick film a good candidate for a tunable device.

Many tunable devices, such as phase shifters [30, 31], varactors [11, 32] and tunable filters [33] have demonstrated the use of thick films. These results, although promising, are not as good as those obtained with thin films in terms of tunability.

c. Thin film: Thicknesses of less than 1 pm are often referred to as thin film and are usually more crystalline in nature, being produced by cathodic sputtering, laser ablation or other thin film processing techniques.

Ferroelectric thin films are very attractive for microwave tunable applications due to their relatively low production cost and, more importantly, their low cost of manufacture,

their low tuning voltage, typically between 2 and 200 volts, depending on the composition of the thin film and its ë thickness.

The dïelectrical response of ieiToeiectric thin films is generally different from that of bulk ferroelectric materials in terms of dieiectric constant and tunability. According to the literature [28, 12], the dielectric constant of thin films is generally lower and the loss tangent, tan 6 is always higher compared to their bulk or thick film counterparts.

Nevertheless, the much lower tuning voltage and other integration advantages still make ferroelectric thin film a very feasible candidate in tunable microwave devices.

The quality of ferroelectric thin films is also highly dependent on the substrates used to deposit the film. Substrates such as magnesium oxide (MgO), alumina or sapphire (Al_2O_3) and lanthanum aluminate ($LaAlO_3$) are often used as substrates in ferroelectric film-based components because of their low loss tangent and good lattice matching.

2.2.4. BST thin film development techniques

a. Ferroelectric thin film

Ferroelectric thin films can be developed using a variety of deposition techniques, each of which has its advantages and disadvantages. Broadly speaking, these techniques can be divided into three groups:

- Physical Vapour Deposition (PVD): this includes RF and magnetron sputtering, molecular jet epitaxy and laser pulse deposition (PLD).
- Chemical Vapour Deposition (CVD): this includes Metal-Organic Chemical Vapour Deposition (MOCVD) and Atomic Layer Deposition (ALD).
- Chemical Solution Deposition (CSD): this includes sol-gel and metal-organic decomposition.

Pulse laser deposition (PLD) has a number of features that make it remarkably competitive in the complex thin-film arena compared with other film growth techniques. In addition, the PLD technique offers simplicity of use, a relatively high deposition rate and is more economical than other deposition techniques.

b. BST thin film

A great deal of work has been reported on the growth of BaxSr1-xTiO3 (BST) ferroelectric thin films.

The use of the PLD technique has ë1.ë reported in recent years [16, 18, 3437]. BST thin films are considered to be 1 of the most promising candidates for tunable microwaves due to their high non-linearity of dielectric constants with the applied electric field and the tunable dielectric properties of Strontium (Sr) and Barium (Ba). BST thin films with x=0.4 - 0.6 are typically used as they demonstrate a large electric field effect at room temperature. The dielectric constant and loss tangent are the two most important parameters affecting the practical applications of BST thin films in tunable microwave devices. These parameters are highly dependent on the quality of the film as well as the substrate used for film deposition [35, 37]. The single-crystal MgO substrate (001) is often chosen for growing BST thin films because of its excellent dielectric properties in terms of low loss tangent, and also, its good durability and low cost compared to other substrates.

2.3. Applications of ferroelectric materials

Ferroelectric materials have received considerable attention over the last decade due to the increasing demand for components with smaller size, lighter weight, higher speed, lower cost and higher power. The change in the dielectric constant of ferroelectric materials with the applied electric field holds the key to a wide range of applications such as varactors, phase shifters and tunable filters.

The next section aims to give an overview of the application of BST ferroelectric thin film in a miniaturised antenna as a dielectric resonator (radiating element).

3. Miniaturisation of a rectangular dielectric resonator antenna by loading a thin film of very high permittivity BST material

One of the important techniques used to achieve miniaturisation is the use of high-permittivity materials. This technique has proved its worth in recent years, particularly for filters and microwave oscillator devices [38]. In addition, for antenna applications, they are also used as single or multilayer substrates [39].

In the literature, a few papers are reported for the miniaturisation of dielectric resonator antennas (DRAs) using very high permittivity materials. For DRAs, these matërials play the role of a radiating element. The advantages of this type of antenna are the variety of shapes available, the different feeding methods (microstrip transmission line, probe feed, coplanar type, etc.) and the absence of metal losses [40].

In recent years, barium strontium titanate (BST) diëlectric material has been widely developed, characterised and integrated into electrical components for a variety of reasons, including the realisation of miniaturisation and/or agility in microwave devices [1]. This type of ceramic material offers high permittivity and low losses that meet the requirements of high-performance microwave systems.

In this section, we propose the miniaturisation of a rectangular dielectric resonator antenna (RDRA) for Wireless Communications Service (WCS).

3.1. Wireless Communication Service (WCS)

According to the FCC, WCS standards are licensed to operate in the 2.3 GHz band and occupy two very narrow frequency bands between 2305-2320 MHz and 2345-2360 MHz (see Figure IV.4) in the RF spectrum range [41] and its use includes mobile phone, text messaging and internet.

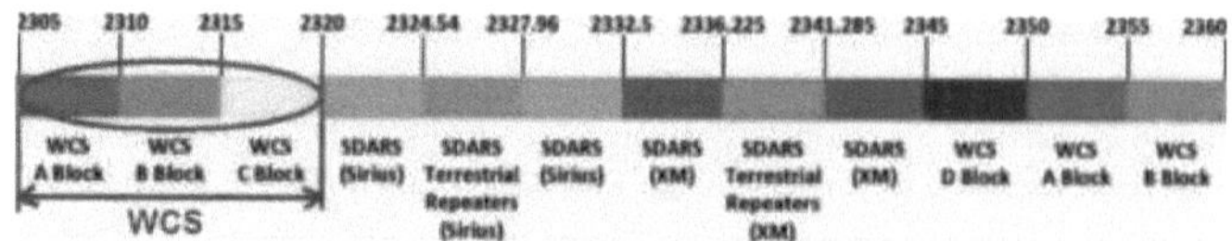

Figure IV.4: Frequency spectrum for WCS and SDARS [41].

3.2. Rectangular Dielectric Resonator Antenna (RDRA) [4]

3.2.1. Basic design (before loading BST thin film)

The basic structure (without an additive layer) consists of a dielectric substrate of Rogers TMM6 material which has a relative permittivity $\varepsilon_{rs} = 6$. The power line and ground plane are printed below and above this substrate, respectively.

A rectangular aperture is cut in the surface of the ground plane. The radiating element formed by Rogers TMM10i dielectric material (with a dielectric constant of 9.8) is mounted diagonally (rotated at an angle a = 45° to the substrate) on the ground plane, and centred over the excitation slot.

Figure IV.5 shows the basic geometry and shape of the RDR antenna.

All the dimensions of the RDR antenna are shown in Table IV.2.

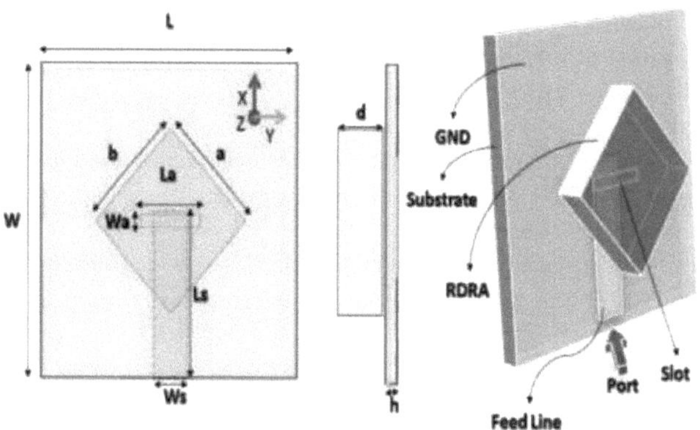

Figure IV.5: Gëomëtrie of the RDR antenna with its basic shape.

Table IV.2: Optimum dimensions of the rectangular RD antenna.

Parameter	Description	Value
L	Substrate length	30 mm
W	Width of substrate	30 mm
h	Substrate height	0.762 mm
ε_{rs}	Substrate material	6 (Rogers TMM6)
a	DR for Length	12.5 mm
b	Width of DR	12.5 mm
d	Height of the DR	6.45 mm
ε_r	DR Materiau	9.8 (Rogers TMM10i)
Ls	Feed length	16 mm
Ws	Feed width	4 mm
La	Slot length	6.95 mm
Wa	Slot width	1.2 mm

3.2.2. **Results and analysis**

The use of the aperture feed mechanism allows excitation by the Transverse Electric (TE_{mnp}) modes of rectangular DRA, where the lowest order mode is TE_{111} [9]. The resonant frequency of this dominant mode can be estimated theoretically using the Dielectric Waveguide Model (DWM), by solving the following transcendental liquidation:

$$k_z \tan(k_z d / 2) = \sqrt{(\varepsilon_r - 1)k_0^2 - k_z^2} \tag{4.9}$$

Or ;
$$k_x^2 + k_y^2 + k_z^2 = \varepsilon_r k_0^2 \quad \text{et} \quad k_0 = 2\pi f_0 \; , \; k_x = m\pi / a \; , \; k_y = n\pi / b \, .$$

f_0 is the free-running frequency and the indices m, n and p represent the variation of the field in the x, y and z directions, respectively.

Using the DWM model, the approximate resonance frequency of the rectangular DRA shown in Figure IV.5 is about 5.7 GHz.

The numerical analysis of this antenna in its original configuration is shown in Figure IV.6 in terms of the reflection coefficient s_{11}.

From the curve in Figure IV.6, it can be seen that the antenna operates around 5.5 GHz and the value of the reflection coefficient at the resonance frequency is less than -30 dB.

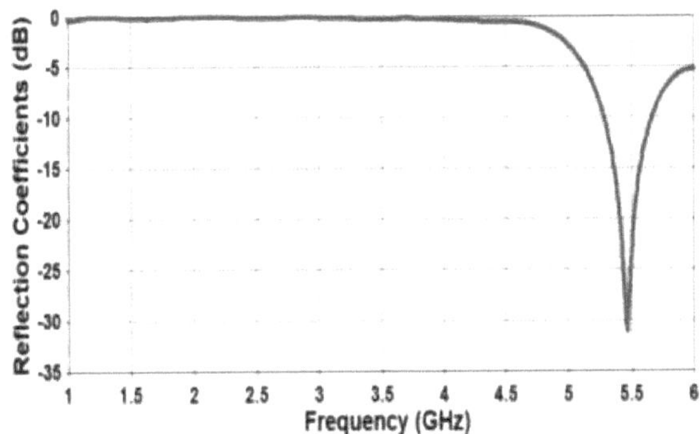

Figure IV.6: Simulated reflection coefficient of the basic RD antenna.

The 3D radiation pattern of the basic RDRA antenna simulated at 5.464 GHz is shown in Figure IV.7.

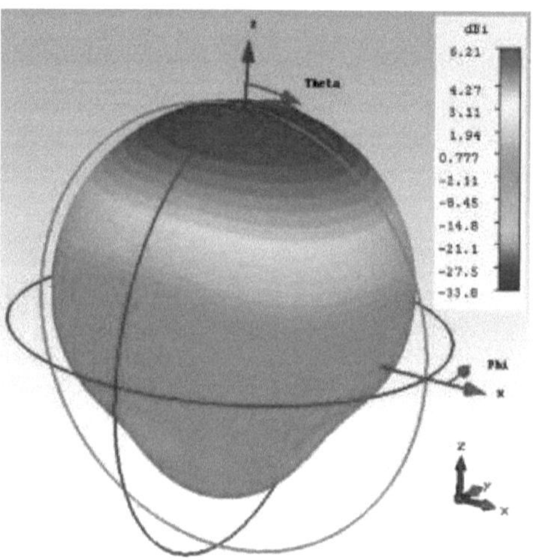

Figure IV.7: 3D radiation pattern of the simulated basic RDRA antenna at 5.464 GHz.

3.3. Miniaturisation process

In this section, we will discuss the miniaturisation procĕdure adoptĕe in this work.

3.3.1. Final design of the antenna at RDR

The same geometry of the antenna structure presented in the prĕcĕdente section is adopted in this case. However, the only change is made to the radiating element. The proposed antenna design consists of two stacked rectangular dieiectric resonator antennas, with width a and length b.

The lower DR (DRi) is made of BST ceramic material with permittivity $s_{ri} = 250$ and height d_i, the upper DR (DR2) has relative permittivity $s_{r2} = 9.8$ (TMM10i) and height d_2, such that: $d_i + d_2 = d = 6.45$ mm, which represents the initial resonator height. Figure IV.8 shows the proposed model incorporating the very high permittivity BST material.

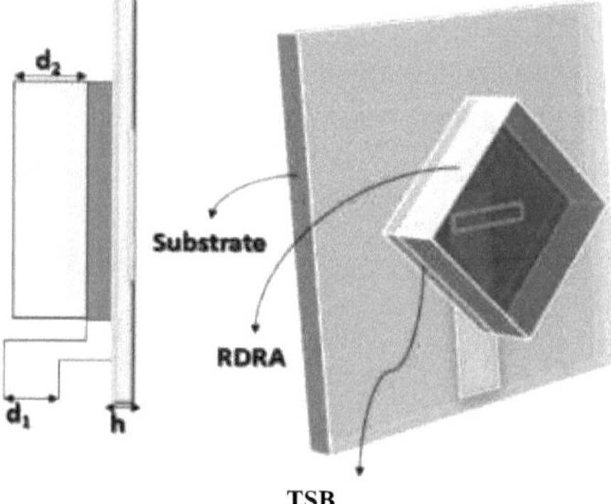

Figure IV.8: Geometry of the RDR antenna integrating the film of BST material.

3.3.2. Results and discussion

a) Reflection coefficient

After including the BST material in the initial structure, the reflection coefficient will be modified. Figure IV.9 shows the simulated reflection coefficient s_{ii} after incorporating the thin film of BST material into the radiating element.

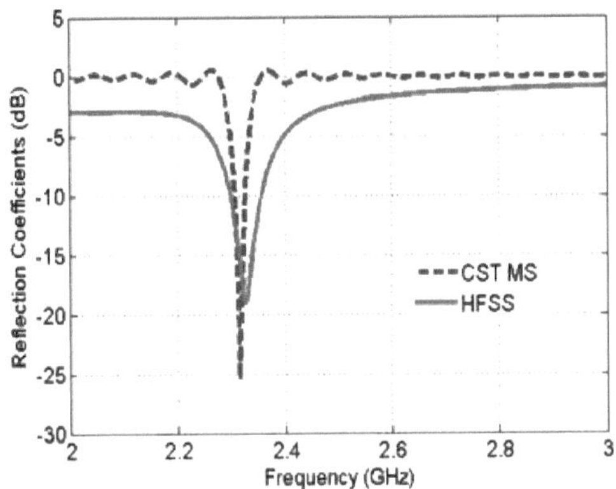

Figure IV.9: Simulated reflection coefficient.

Simulations were carried out using both CST Microwave Studio and Ansoft HFSS software. The reflection coefficients, represented in Figure IV.9, are in good agreement. From these curves, it can be seen that the proposed antenna has a resonant frequency around 2.314 GHz with an amplitude reflection coefficient between -20 dB and -25 dB. In addition, the proposed design offers an impëdance bandwidth (for a reflection coefficient below -10 dB) of 20 MHz.

b) Miniaturisation rate

The area occupied by the diëlectrical resonator is a crucial factor in assessing the rate of miniaturisation. For this reason, the DWM model is used to estimate the equivalent diëlectric resonator antenna that resonates at the same frequency as our final ADR empire. After calculation, it is dëmontrë that for the same diëlectrical permittivity^ (9.8) and height (0.645 mm), the size of the resonator must be of the order of 42.5mm x 42.5mm to operate around the 2.31 GHz frequency.

Figure IV. 10 shows the size of the resonator antenna with (the rectangle in dark colour) and without (the rectangle in light colour) loading of the BST ceramic layer.

Comparing the two rectangles, it is clear that the area occupied by the resonator with BST represents 10% of the size of the resonator without integrating this layer. This shows the importance of the miniaturisation technique adopted.

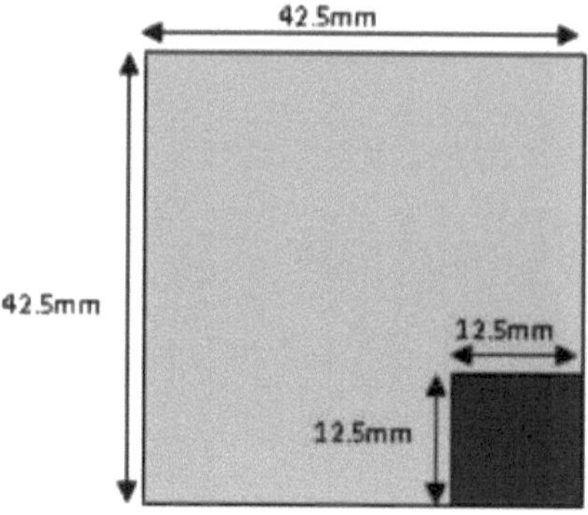

Figure IV.10: Comparison between the resonator size with and without loading of the thin layer of BST matëriau.

c) Radiation diagrams

The simulated results of the 2D and 3D radiation patterns at the resonant frequency of 2.314 GHz are calculated in the two main planes (XZ and YZ plane), as shown in Figures IV.11 and IV.12, respectively. From these figures, it can be seen that the proposed antenna presents a bidirectional radiation pattern in the XZ plane and an omnidirectional behaviour in the YZ plane.

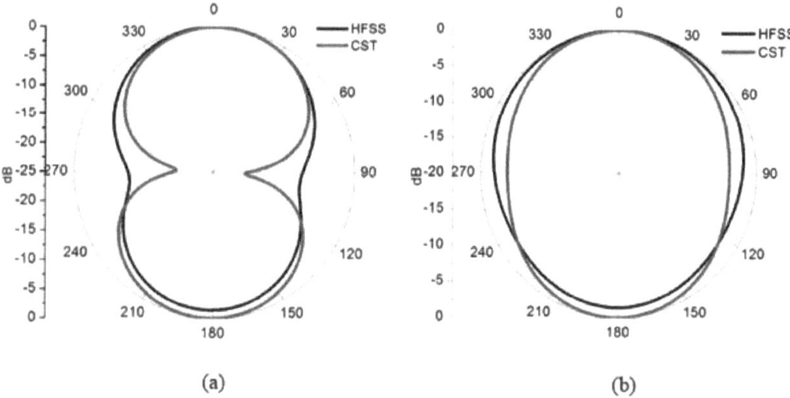

Figure IV.11: Simulated radiation patterns at 2.314 GHz in the; (a) XZ plane, (b) YZ plane.

94

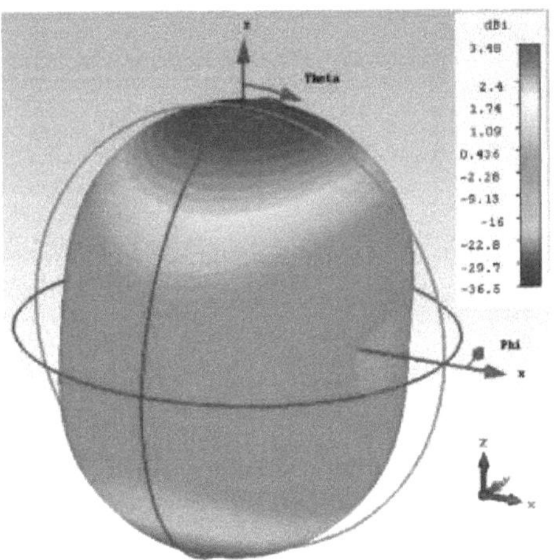

Figure IV.12: 3D radiation pattern of the RDRA antenna loaded by the BST layer simulated at 2.314 GHz by the CST software.

3.4. Parametric studies and discussion

In order to analyse the effect of the antenna structure parameters on its performance, a parametric study is carried out by acting on certain physical properties of the antenna to show the best optimisation result. The CST software was used to study the effect of the parameters shown in Table IV.2 on the performance of the RDRA antenna.

3.4.1. Effect of dielectric resonator parameters

The performance of the proposed antenna can be affected by the geometries of the rectangular RD and its permittivity. In this section, we will discuss the effects of the parameters; the diëlectrical permittivity $r_{,r}2$ of the thin film of the BST material, the heights d_i and D and the two dimensions (length a and width b) of the rectangular resonator on the reflection coefficient.

a. Effect of the dielectric permittivity ε_{r2} of the BST film

Figure IV. 13 shows the effect of the s_{r2} permittivity of the BST layer on the reflection coefficient, keeping its thickness constant (d1 = 0.645 mm). The permittivity values are chosen between 9.8 (this is the s_{ri} permittivity of the TMM10i material) and 350.

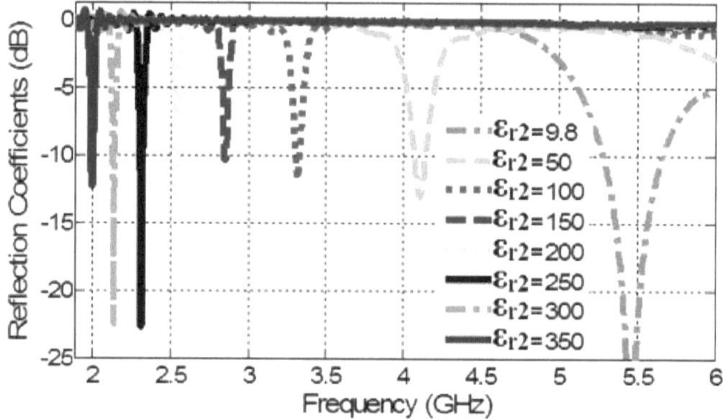

Figure IV.13: Effect of permittivity <'m on the reflection coefficient.

From the curves in this figure, we can see that increasing the permittivity of the BST film causes a decrease in the resonant frequency of the RDRA antenna and the associated bandwidth becomes small.

The variation in resonance frequency as a function of permittivity is shown in Figure IV. 14, where an exponential degradation of the resonance frequency is observed as a function of the increase in permittivity S_{r2}.

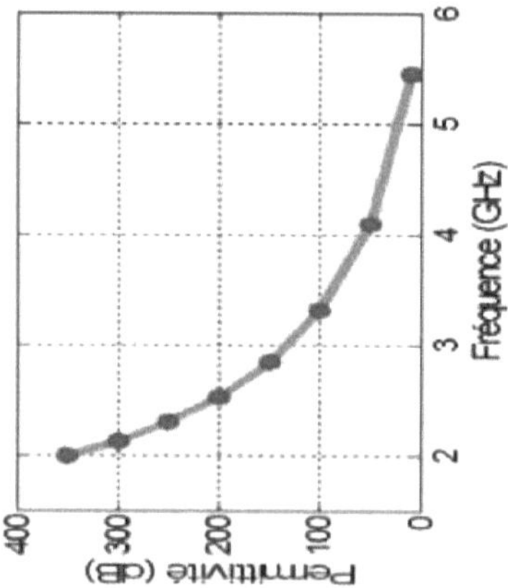

Figure IV.14: Effect of permittivity on the rysonance frequency of the proposye antenna.

b. Effect of film thickness d_i BST

Figure IV.15 shows the variation of the reflection coefficient as a function of the thickness of the thin layer of BST material.

From the curves in this figure, it can be seen that the thickness d_i of the very high permittivity BST film has a large effect on the reflection coefficient of the proposed antenna proposëe.

Increasing the BST thickness by a very small value (100 цт) causes the resonance frequency of the RDRA antenna to shift to the left.

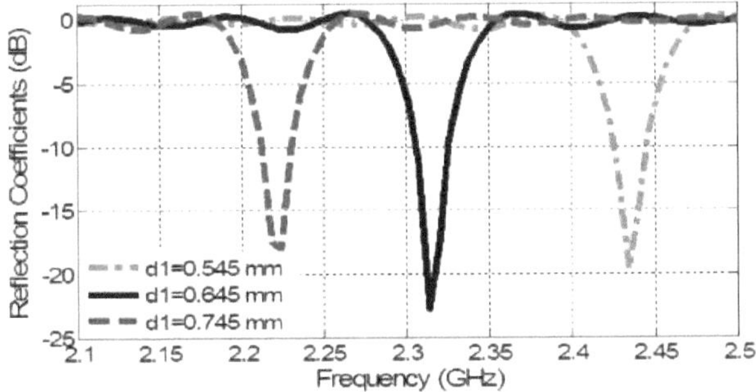

Figure IV.15: Effect of parameter d_i on the reflection coefficient.

The operating band of the proposed antenna (wireless communication services; WCS) is lost in front of the shift which corresponds to $d_1=0.645$ mm to the left or to the right of the PIC of the resonance frequency.

c. Effect of TMM10i resonator geometries (a, b, d, a)

The effects of the dimensions; length a, width b and height d of the TMM10i rectangular resonator and its angular variation a on the reflection coefficient of the RDRA antenna are shown in Figures IV.16, IV.17, IV.18 and IV.19, respectively.

Unlike the BST thin film, the variation -increasing or decreasing- by one millimetre in the three dimensions (a, b, d) of the TMM10i radiating element, does not have a great effect on the antenna's reflection coefficient.

This is due to its low permittivity compared with the high permittivity of the BST material, which has a considerable effect on the resonance frequency.

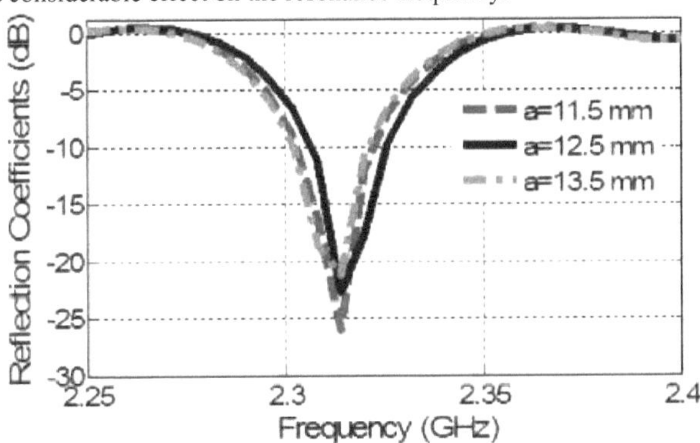

Figure IV.16: Effect of the length a of the RDR on the reflection coefficient.

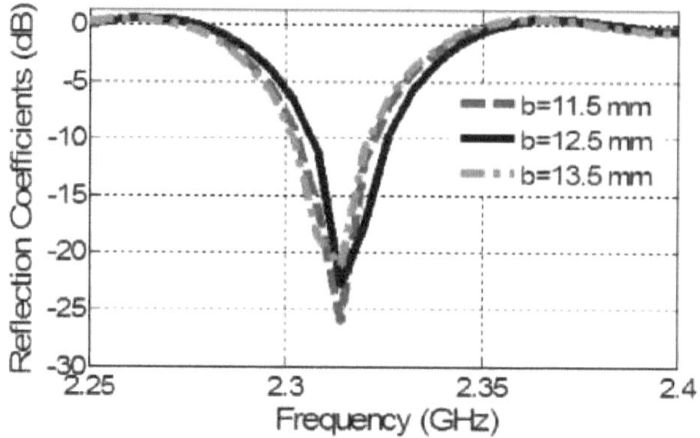

Figure IV.17: Effect of the width b of the RDR on the reflection coefficient.

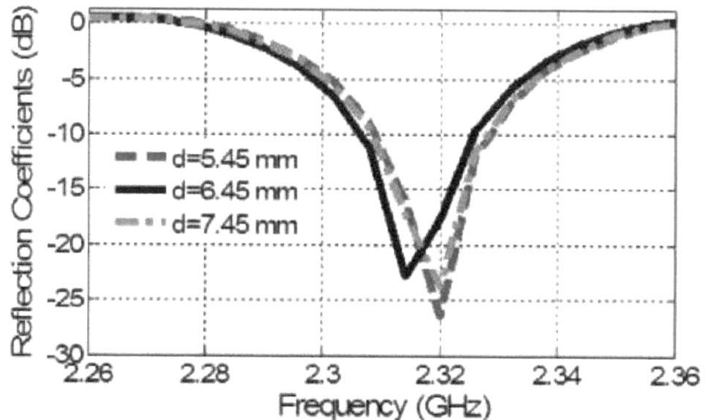

Figure IV.18: Effect of the height d of the RDR on the reflection coefficient.

The same comment applies to parameter a (the angle of rotation of the RD relative to the substrate) concerning its poor effect on the reflection coefficient.

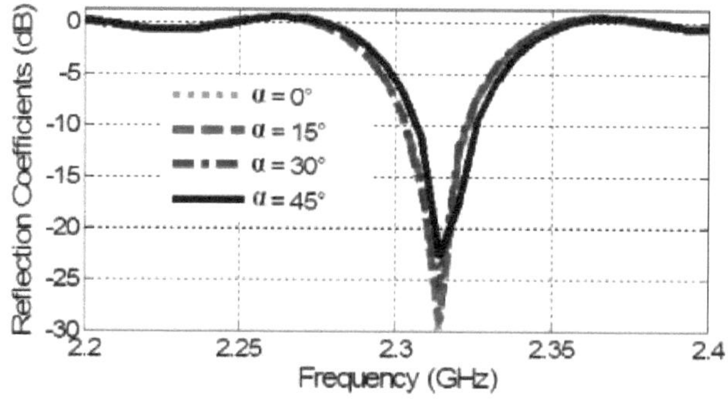

Figure IV.19: Effect of varying the angle a on the reflection coefficient.

3.4.2. Effect of feed line geometries

The effects of feed line length and width (parameters Ls and w_s) on the antenna reflection coefficient are shown in Figures IV.20 and IV.21. From figure IV.20, it can be seen that a small variation (0.5 mm) in the Ls parameter causes a dëgradation in the value of the reflection coefficient. The best adaptation is recorded for the case where $L_{s=16}$ mm.

A 2 mm reduction in the w_s parameter (as shown in Figure IV.21) allows the RDRA antenna to be miniaturised (shifted towards the lowest frequency spectrum).

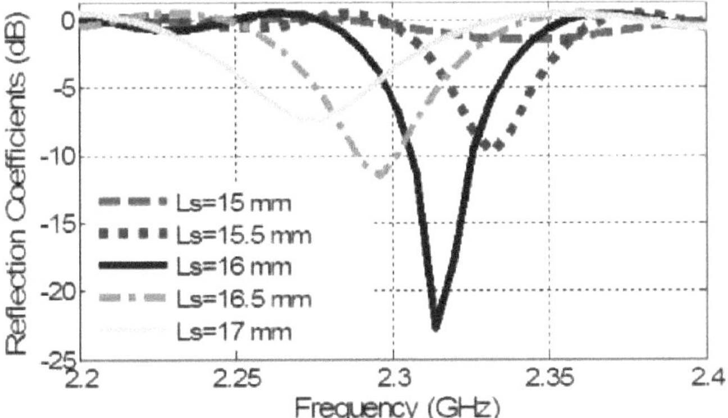

Figure IV.20: Effect of parameter Ls on the reflection coefficient.

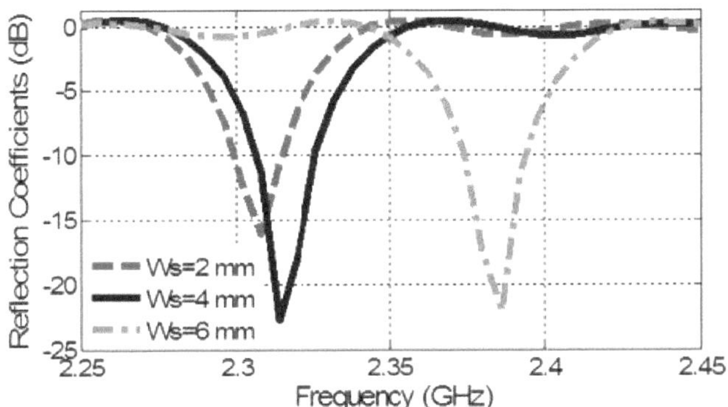

Figure IV.21: Effect of the parameter Ws on the reflection coefficient.

3.4.3. Effect of the geometry of the aperture (the slot)

Figures IV.22 and IV.23 show the effect of the length and width of the slit, gravée in the ground plane, on the reflection coefficient. It can be concluded from these curves that the parameter L_a has a considërable effect on the reflection coefficient (Figure IV. 22). However, the resonance frequency is slightly dëcalée for the w_a parameter (Figure IV.23).

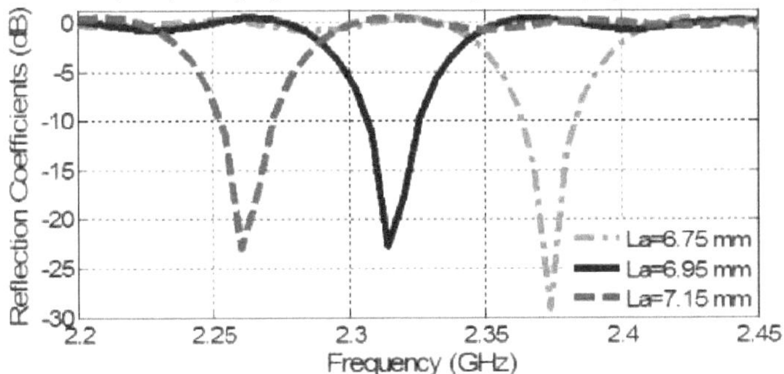

Figure IV.22: Effect of parameter La on the reflection coefficient.

From the ten figures (between Figures IV.13 and IV.23), it can be concluded that the optimum values of the parameters studied, which guarantee that the DRA antenna covers the desirable band (wireless communication service), are those mentioned in Table IV.2.

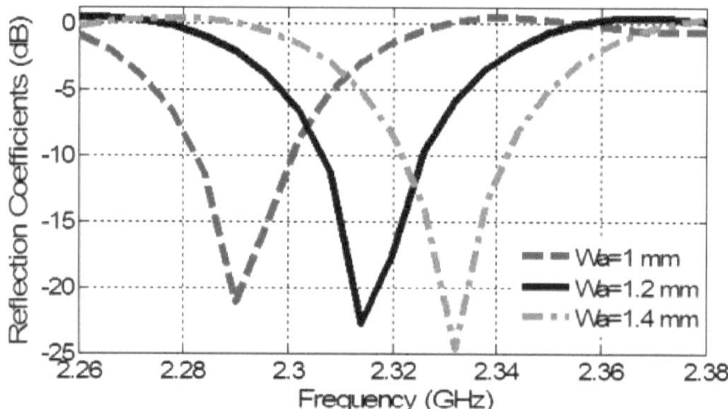

Figure IV.23: Effect of the Wa parameter on the reflection coefficient.

4. Conclusion

In this chapter, a new miniaturisëe electric dië resonator antenna has ëlëtudiëe numëriquement and analysëe using two ëlectromagnëtiques simulators. The use of a high permittivity diëlectric material (BST ceramic material) allows this miniaturisation to be achieved. With this technique, the reduction in radiated radiation reaches 90%.

The results obtained show that the proposed design has a resonant frequency of 2.314 GHz and offers a bandwidth between 2.306 GHz and 2.325 GHz. With these radiation characteristics, the proposed RD antenna can be a suitable candidate for wireless communication service (WCS) systems.

It is desirable to provide an expërimental ëtude to support our work; and this necessitates completing two phases:

The first phase involves manufacturing the RDRA antenna and the thin film of BST material;

- A specific method for producing ferroelectric material.
- Prepare the barium-strontium titanate components (the ë samples: Ba, Sr, TiO_3).
- Use the pu^ laser dëpôt (PLD) technique to dëvelop a ëthickness of $d_1=0.645$ mm of BST film.
- Cara^riser the film to measure the value of its relative diëlectrical permittiv^.
- Prepare the two materials with the heights indicated in table IV.2: TMM6i for the substrate, and TMM10i for the diëlectrical resonator.
- Etch the substrate or the printed circuit on both sides (the ground plane and the power supply line) using the photo-etching method.
- Using a laser machine, cut a ëthickness $d_i=0.645$ mm from the TMM10i matëriau.
- Glue the BST film to the bottom surface of the TMM10i (in place of the coupëe piece).
- Place both resonators diagonally above the opening in the ground plan.

And the second phase concerns the Slectrical and SlectromagnStic measurements before and after loading the RDRA antenna with the BST matëriau;

- Measurement of rSflection coefficient.
- Measurement of the radiation pattern, efficiency and gain of the RDRA antenna.
Experimental results are recommended in the future, to concretise the theoretical principle and the design of the RDRA antenna loaded by the BST material.

Bibliography of Chapter IV

[1]   D. Remiens, F. Ponchel, A. Ghalem, T. Lasri. *"Ferroelectric thin films working at microwave frequency for reconfigurable devices: performances comparison of BST, PST,* " International Journal of Materials Engineering Innovation. 2014, Vol. 5, no 4, pp. 327-335, 2014.

[2]   F.H. Wee and F. Malek , *"Gain Enhancement of a Microstrip Patch Antenna using Array Rectangular Barium Strontium Titanate(BST),"* The 2011 Loughborough Antennas and Propagation Conference, November 14-15, 2011, Loughborough, UK.

[3]   PhD thesis in electronic, electrical and computer engineering, Xiang Gao, ' *Antenna Designs Based On Metamaterial-Inspired Structures*', University of Birmingham, September 2016.

[4]   Farouk Chetouah, Nacerdine Bouzit, Idris Messaoudene, Salih Aidel, Massinissa Belazzoug, Boualem Hammache, '*Miniaturized rectangular dielectric resonator antenna for WCS*', 12th International Conference on Innovations in Information Technology (IIT-2016), Pp. 1-4, 28-30 Nov. 2016, Al-Ain, United Arab Emirates.

[5]   D. thesis in Electrical and Electronic Engineering, Shaozhen Zhu, '*Wearable antennas for Personal wireless Networks*', University of Sheffield, January 2008.

[6]   D. thesis in Electrical and Electronic Engineering, Shahid Bashir, '*Design and Synthesis of Non Uniform High Impedance Surface based Wearable Antennas*', Loughborough University, October 2009.

[7]   Yi Huang, Kevin Boyle, '*Antennas From Theory To Practice*', Page 14, John Wiley & Sons Ltd, 2008.

[8]   F. H. Wee; F. Malek, '*Barium Strontium Titanate (BST) array antenna covered with dielectric resonator superstrates for high gain and high directive antenna*', 9th International Symposium on Antennas Propagation and EM Theory (ISAPE2010), Pp. 112 - 115, 29 Nov.-2 Dec. 2010, Guangzhou, China.

[9]   D. in Electronic, Electrical and Computer Engineering, Leong Ching Cheng, '*Ferroelectric Microwave Circuits*', University of Birmingham, June 2009.

[10]   M. J. Lancaster, J. Powell, and A. Porch, *"Thin-film Ferroelectric Microwave Devices,"* Superconductor Science & Technology, vol. 11, pp. 1323-1334, Nov 1998.

[11]   R. Jakoby, P. Scheele, S. Muller, and C. Weil, *"Nonlinear Dielectrics for Tunable Microwave Components,"* 15th International Conference on Microwaves, Radar and Wireless Communications, vol. 2, p. 369, 2004.

[12]   A. K. Tagantsev, V. O. Sherman, K. F. Astafiev, J. Venkatesh, and N. Setter, *"Ferroelectric Materials for Microwave Tunable Applications,"* Journal of Electroceramics, vol. 11, pp. 5-66, 2003.

[13]   M. Kamlah, *"Ferroelectric and Ferroelastic Piezoceramics - Modeling of Electromechanical Hysteresis Phenomena,"* Continuum Mechanics and Thermodynamics, vol. 13, p. 219, 2001.

[14]   T. Remmel, R. Gregory, and B. Baumert, *"Characterization of Barium Strontium Titanate Films Using XRD,"* International Centre for Diffraction Data, 1999.

[15]   B. Acikel, T. R. Taylor, P. J. Hansen, J. S. Speck, and R. A. York, *"A New High Performance Phase Shifter Using BaxSr1-xTiO3Thin Films,"* IEEE Microwave and Wireless Components Letters, vol. 12, pp. 237-239, 2002.

[16]   B. J. Kim, S. Baik, Y. Poplavko, Y. Prokopenko, J. Y. Lim, and B. M. Kim, *"Epitaxial BSTO Thin Films for Microwave Phase Shifters,"* Asia-Pacific Microwave Conference, pp. 934-937, 2000.

[17]   J. B. L. Rao, D. P. Patel, L. C. Sengupta, and J. Synowezynski, *"Ferroelectric Materials for Phased Array Applications,"* IEEE 1997 Antennas and Propagation Society International Symposium Digest, vol. 4, pp. 2284-2287 vol.4, 1997.

[18]   F. W. Van Keuls, C. T. Chevalier, F. A. Miranda, C. M. Carlson, T. V. Rivkin, P. A. Parilla, J. D. Perkins, and D. S. Ginley, *"Comparison of the Experimental Performance of Ferroelectric CPW Circuits with Method-of-Moment Simulations and Conformal Mapping Analysis,"* Microwave and Optical Technology Letters, vol. 29, pp. 34-37, 2001.

[19] P. M. Suherman, T. J. Jackson, Y. Y. Tse, I. P. Jones, R. I. Chakalova, M. J. Lancaster, and A. Porch, "*Microwave Properties of Ba0.5Sr0.5TiO3 Thin Film Coplanar Phase Shifters,*" Journal of Applied Physics, vol. 99, p. 104101, 2006.

[20] N. M. Alford, S. J. Penn, A. Templeton, X. Wang, J. C. Gallop, N. Klein, C. Zuccaro, and P. Filhol, "*Microwave Dielectrics,*" IEE Colloquium on Electro-technical Ceramics - Processing, Properties and Applications, pp. 9/1-9/5, 1997.

[21] S. J. Penn, N. McNalford, A. Templeton, N. Klein, J. C. Gallop, P. Filhol, and X. Wang, "*Low Loss Ceramic Dielectrics for Microwave Filters,*" IEE Colloquium on Advances in Passive Microwave Components, pp. 6/1-6/6, 1997.

[22] S. S. Gevorgian, E. F. Carlsson, S. Rudner, U. Helmersson, E. L. Kollberg, E. Wikborg, and O. G. Vendik, "*HTS/Ferroelectric Devices for Microwave Applications,*" IEEE Transactions on Applied Superconductivity, vol. 7, pp. 2458-2461, 1997.

[23] O. G. Vendik, E. Kollberg, S. S. Gevorgian, A. B. Kozyrev, and O. I. Soldatenkov, "*1 GHz Tunable Resonator on Bulk Single Crystal STO Plated with YBCO Films,*" Electronics Letters, vol. 31, pp. 654-656, 1995.

[24] T. Chakraborty, I. Hunter, R. Kurchania, A. A. B. A. Bell, and S. Chakraborty, "*Intermodulation Distortion in Wide-and Dual-mode Bulk Ferroelectric Bandpass Filters,*" IEEE MTT-S International Microwave Symposium Digest, p. 4 pp., 2005.

[25] M. K. Roy, C. Kalmar, R. R. Neurgaonkar, J. R. Oliver, and D. Dewing, "*A Highly Tunable Radio Frequency Filter Using Bulk Ferroelectric Materials,*" 14th International Symposium on Applications of Ferroelectrics IEEE, p. 25, 2004.

[26] D. M. Kosmin, V. N. Osadchy, and A. B. Kozyrev, "*Switching Time of Bulk Ferroelectric Sandwich Varactors,*" 17th International Crimean Conference Microwave & Telecommunication Technology, p. 533, 2007.

[27] J. B. L. Rao, D. P. Patel, and V. Krichevsky, "*Voltage-Controlled Ferroelectric Lens Phased Arrays,*" IEEE Transactions on Antennas and Propagation, vol. 47, p. 458, 1999.

[28] O. G. Vendik, E. K. Hollmann, A. B. Kozyrev, and A. M. Prudan, "*Ferroelectric Tuning of Planar and Bulk Microwave Devices,*" Journal of Superconductivity, vol. 12, pp. 325-338, 1999.

[29] L. Sengupta and S. Sengupta, "*Novel Ferroelectric Materials for Phased Array Antennas,*" IEEE Transactions on Ultrasonics, Ferroelectrics and Frequency Control, vol. 44, pp. 792797, 1997.

[30] K. S. K. Yeo, W. F. Hu, M. J. Lancaster, B. Su, and T. W. Button, "*Thick Film Ferroelectric Phase Shifters Using Screen Printing Technology,*" 34th European Microwave Conference, 2004, pp. 1489-1492.

[31] W. Hu, D. Zhang, M. J. Lancaster, K. S. K. Yeo, T. W. Button, and B. Su, "*Cost Effective Ferroelectric Thick Film Phase Shifter based on Screen-Printing Technology,*" IEEE MTT- S International Microwave Symposium Digest, pp. 591-594, 2003.

[32] W. Hu, D. Zhang, M. J. Lancaster, T. W. A. B. T. W. Button, and B. A. S. B. Su, "*Investigation of Ferroelectric Thick-Film Varactors for Microwave Phase Shifters,*" Microwave Theory and Techniques, IEEE Transactions on, vol. 55, pp. 418-424, 2007.

[33] P. Scheele, S. Muller, C. Weil, and R. Jakoby, "*Phase-shifting Coplanar Stubline-Filter on Ferroelectric-Thick Film,*" 34th European Microwave Conference, vol. 3, pp. 1501-1504, 2004.

[34] W. T. Chang, S. W. Kirchoefer, J. A. Bellotti, and J. M. Pond, "*(Ba,Sr)TiO3Ferroelectric Thin Films for Tunable Microwave Applications,*" Revista Mexicana De Fisica, vol. 50, pp. 501505, 2004.

[35] C. L. Chen, J. Shen, S. Y. Chen, G. P. Luo, C. W. Chu, F. A. Miranda, F. W. V. Keuls, J. C. Jiang, E. I. Meletis, and H. Y. Chang, "*Epitaxial Growth of Dielectric BSTO Thin Film on MgO for Room Temperature Microwave Phase Shifters,*" Applied Physics Letters, vol. 78, pp. 652-654, 2001.

[36] J. Kim, I. K. Yu, S. J. Lee, and K. Y. Kang, "*Growth and Characterization of BSTO and YBCO/BSTO Thin Films on MgO (100) Substrates,*" Journal of the Korean Physical Society, vol. 32, pp. 183-185, 1998.

[37] J. Zhang, H. Zhang, K. J. Chen, S. G. Lu, and Z. Xu, "*Microwave Performance Dependence*

of BST Thin Film Planar Interdigitated Varactors on Different substrates," in 2nd IEEE International Conference on Nano/Micro Engineered and Molecular Systems, 2007, pp. 678-682.

[38] S. J. Fiedziuszko, I. C. Hunter, T. Itoh, Y. Kobayashi, T. Nishikawa, S. N. Stitzer, and K. Wakino, "*Dielectric Materials, Devices, and Circuits*", IEEE Transactions on Microwave Theory and Techniques, Vol.50,No. 3,706-720,2002.

[39] J. M. Laheurte, L .P. B. Katehi, G. M Rebeiz, "*CPW-fed slot antennas on multilayer dielectric substrates,*" IEEE Transactions on Antennas and Propagation, Vol.44,No. 8,1102-1111, 1996.

[40] K. M. Luk and K. M. Leung, "*Dielectric Resonator Antennas*," Research Studies Press LTD. England, 2003.

[41]https://www.fcc.gov/wireless/bureau-divisions/broadband-division/wireless-communications-service-wcs

Chapter V

Numerical and experimental analysis of miniature microstrip antennas

1. Introduction

Active and passive microwave components ийНзёз in the production of modern individual tëlëcommunication devices are experiencing significant and gradual progress in the field of their size miniaturization. However, the antenna surface, considered as the main element for signal transmission and reception, still occupies the largest volume in wireless communication systems. Reducing the size of the antenna increases its resonance frequency and its bandwidth becomes ёйюНс and vice versa.

Several ëtudes have ёlё rëalisëes so far for the miniaturisation of antennas; among these ëtudes is the ground plane default structure (DGS) technique. Several forms of ground plane have ёlё ëtudiëed in the scientific literature with different DGS structures.

For the monopole printed antenna, the work consists of modifying the excitation lines as well as reducing the surface area of the ground plane to create an inhomogënëitë in the movement of the ëlectromagnëtic waves.

In this chapter, two microstrip antennas are fabricated and testedë for wireless local area network (WLAN) applications. The study of the first antenna with narrow bandwidth will be discussedëe in section 4, while the other one with wide bandwidth will be presented in section 5. The DGS technique is employed, the aim of which is to achieve miniaturisation of microstrip antennas; the ground plane structure used has an 'L-inverse' shape.

At the end of sections 4 and 5, a complete paramëtrical study is made to look for the best performance of the microstrip antenna.

2. Wireless local area network (WLAN)

Following a dëcision made in 1985 by the US Federal Communications Commission (FCC), 802.11 or Wi-Fi has ёl.ё come to light by opening up several wireless spectrum bands.

IEEE 802.11 is a standard protocol for wireless local area network (WLAN) systems. The IEEE 802.11a standard considers the frequency spectrums; 5.15-5.35 GHz and 5.725-5.825 GHz as the ëtransmitter-rëceiver band and receiver band, respectively. In addition, the IEEE 802.11bg standard is adëquate for 2.4 GHz (2.4-2.484 GHz) for wireless local area networks.

IEEE 802.11a operates in the available 5 GHz bands and employs the most ëlevëes frequency bands that are predominantly usedëes in the network, due to its more ёkyë cost.

3. Production and measurement procedure

3.1. Procedure for manufacturing printed circuits using the photogravure method

The microstrip antennas manufactured in the work in this chapter are double-sided printed circuits whose substrate is a dielectric FR4 material.

Among the methods used to manufacture printed circuits is the photogravure method, which is based on four essential stages:

a) Printing layout diagrams

For the CST Software interface, the two antenna design plans are exported to Autocad software in the 2D form of a file with a .DXF extension (figure V.1) and printed (with a 1x1 format) on a transparent sheet (two masks).

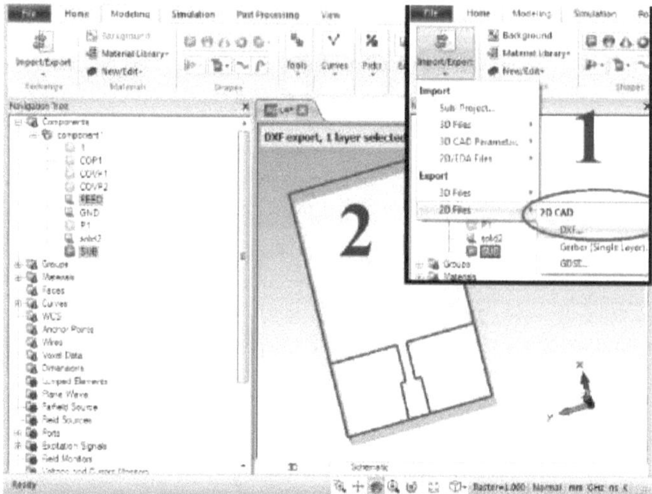

Figure V.1: Exporting a DXF extension file through the CST software interface.

b) Insulation of the double-sided printed circuit protected by the two masks with UV rays (Figure V.2).

c) Revelation of the epoxy plate exposed to light in a NaOH solution to be developed.

d) Etching of the plate in an iron perchloride bath, which removes the unprotected copper.

Figure V.2: UV isolator for printed circuits.

e) 1.1 Choice of substrate

The substrate is a matĕrial support for the microstrip antenna, among the criteria for choosing a substrate may be mentioned:

- The development technique used (photogravure, laser engraving, printing on bare copper, etc.).
- The values of the diĕlectrical constant and the loss.
- Thickness h.
- Cost and availability.

f) 1.2 SMA connector

The SMA (for: SubMiniature version A) connector is widely used to characterise microwave devices. It is a coaxial connector with a characteristic impedance of 50 Q (see figure V.3).

Figure V.3: SMA connectors. [1,2]

3.2. Measurement equipment used

3.2.1. Vector network analyser

To measure the dispersion parameters (for example, the reflection coefficient, s_{11}) of microstrip antennas, we used two different types of vector network analyser:

a) The N5224A is available from the CDTA centre (Centre de Dëveloppement des Technologies Avancëes) in Baba Hassen, Algiers.

b) The Agilent 8719ES at the EMP school (Ecole Militaire Polytechnique) in Bordj El Bahri, Algiers.

c) PNA N5224A network analyser [3]

The PNA N5224A (Figure V.4) high performance Keysight technology measurement instrument is a vector network analyser comprising an integrated S-parameter test set, SSD drive, mouse, keyboard (US style), USB interfaces and a 10.4" LCD touch screen. The N5224A has robust 2.4 ohm to 50 ohm test ports. The N5224A has a sweep frequency range from 10 MHz up to 43.5 GHz, which is more than sufficient for the microwave applications in our study.

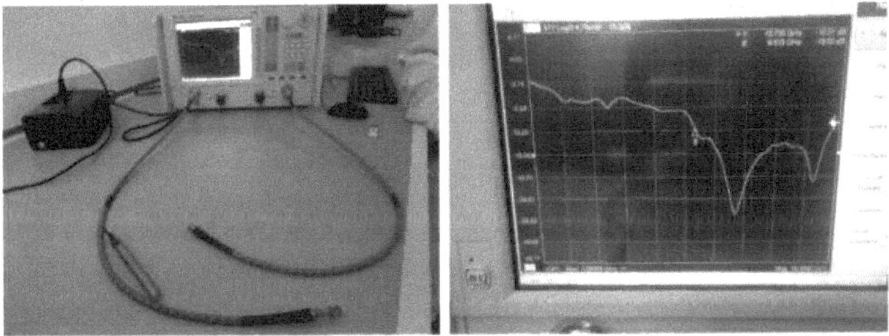

Figure V.4: The PNA N5224A network analyser.

d) Agilent 8719ES Network Analyzer [4]

The Keysight Agilent 8720ES Vector Network Analyser (Figure V.5) provides comprehensive characterisation of RF and microwave components. The Agilent 8720ES comprises an integrated synthesised source, test set and tuned receiver. The integrated S-parameter test set provides a full range of modulus and phase measurements in both forward and reverse directions.

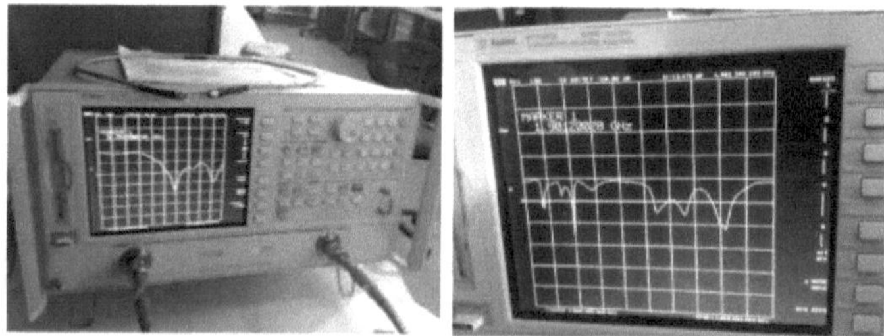

Figure V.5: The Agilent 8719ES network analyser.

e) Technical characteristics of the N5224A and 8719ES analysers

The technical characteristics of the two network analysers N5224A and 8719ES are shown in Table V. 1.

Table V.1: Technical characteristics of the N5224A and 8719ES network analysers [5,6].

Features	N5224A	8719ES
Minimum/maximum frequency	10 MHz/43.5 GHz	50 MHz/13.5 GHz
Dynamic range	127 dB	100 dB
Output power	13 dBm	10 dBm
Noise trace	0.003 dBrms	-
Number of integrated ports	2 or 4 ports	2 ports
Harmonics	-60 dBc	-
Noise level	-114 dBm	-
Best speed at 201 points, 1 sweep	5.5 ms	-
Applications	S parameters Equilibrium measurements Materials RF pulse	S parameters RF pulse balance measurements
Components	Antennas Mixer / frequency converter Amplifiers	Antennas Amplifiers
Touch screen	Yes	No

f) Calibration of the N5224A analyser

With precise calibration in advance in the frequency range configuration, the network analyser can provide graphical results in terms of the s_{11} reflection coefficient of the antenna under test as a function of frequency. The network analyser can also provide the measurement s_{21} using one of its ports as transmission and the other as reflection.

Calibration of the N5224A vector network analyser is carried out automatically by the N4691B electronic calibration module (see Figure V.6).

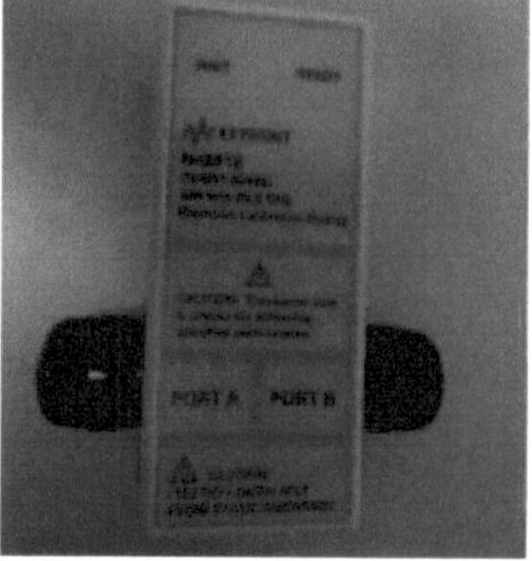
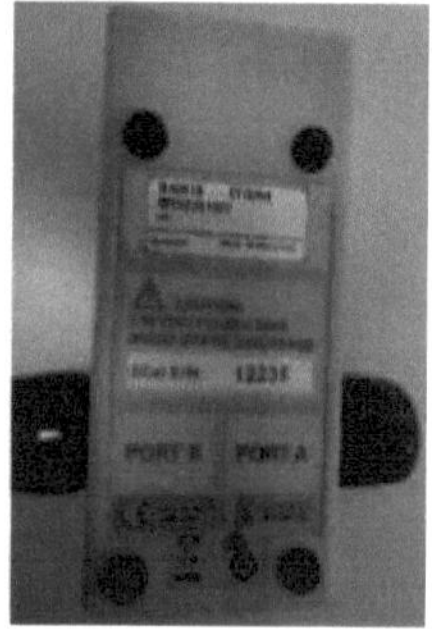

Figure V.6: N4691B electronic calibration module.

The N4691B module must be connected between two ports (A and B) of the network analyser via two HF probes, as shown in Figure V.7. After configuring the frequency range and the number of iterations used during antenna measurements, the operation can be started:
-       A 'WAIT' LED lights up during the calibration test and a faster scan is applied to the reflection standards (OPEN, SHORT, LOAD) connected to the source port to be calibrated.
-       After a few minutes (between 3 and 5 minutes), calibration ends with the 'READY' LED lighting up.

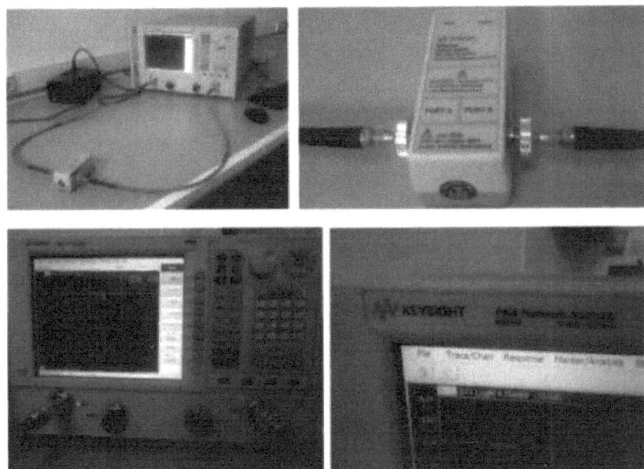

Figure V.7: Calibration of the N5224A analyser by the N4691B module.

3.2.2. The anechoic chamber [7]

In this chapter, the radiation characteristics of the microstrip antenna were ë1.ë measured in an electromagnetic anechoic chamber using a far-field antenna measurement system.
Figure V.8 shows an interior view of the anechoic chamber in the microwave laboratory at the Institut Nationale de la Recherche Scientifique (INRS), Montreal - Canada.

Figure V.8: Interior view of an anechoic chamber at INRS. [8]

The anechoic chambers developed for outdoor testing are used to characterise the antenna's radiation performance. The test is carried out inside a chamber whose walls are lined with an RF absorber to minimise electromagnetic interference.

The configuration shown in Figure V.9 comprises a transmit antenna, an azimuth turntable and an antenna mounting box. A standard gain linearly polarised horn antenna (working frequency range: 900 MHz-18 GHz) has been used as the transmit antenna. The antenna under test is mounted on the turntable in the remote range of the transmitter. During the measurement, the turntable is automatically controlled by the software to rotate through 360° (via a stepper motor) and the power level received is recorded as a function of the phase of rotation.

It is important to note that the location of the measured antenna must be aligned with the centre of the turntable, so that the test results can be symmetrical. The measured data is recorded by the network analyser as a function of frequency and azimuth angle. The data is then transferred to the computer and plotted as 2D radiation patterns. The 2D traces are taken for the Theta (θ) and Phi (φ) slices for vertical and horizontal polarisations in the azimuth plane in this study.

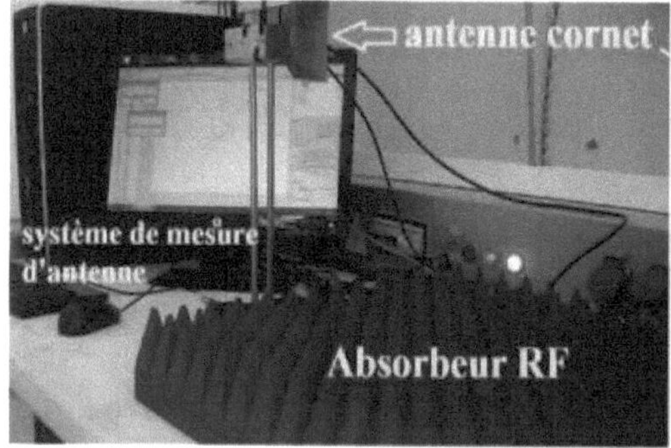

(a)

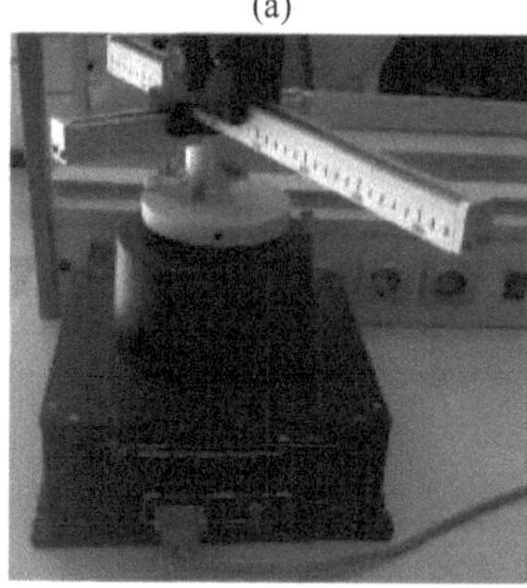

(b)

Figure V.9:(a) Chamber configuration for radiation pattern measurement, (b) Rotary table.

4. Miniaturised narrow-band monopole antenna [9]

4.1. Design configuration

The basic antenna structure consists of a rectangular patch excité by a microstrip transmission line and тртшё on an FR-4 substrate (;:,■ = 4.4, tan5 = 0.02), while the ground plane is тртё on the other sideё of the diёlectric substrate, as illustrated^ in Figure V.10.a. In order to achieve miniaturisation of the antenna, we ёtudiё the same antenna structure, with a partially imprimё ground plane and with a slot in the excitation line, as illust^ed in Figure V.10 (b and c) and Figure V.10.d, respectively. The details of the four cases are as follows:

- **Case 1**: a rectangular monopole antenna with a complete ground plan.
- **Case 2**: a rectangular monopole antenna with a printed half-ground plane.
- **Case 3**: a rectangular monopole antenna with a printed ground plane of inverted L shape.
- **Case 4**: a rectangular monopole antenna with a slot on the excitation linc.

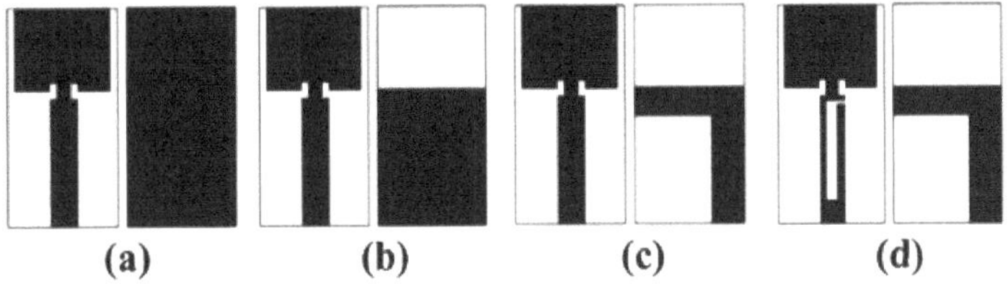

Figure V.10: The evolution of the antenna design. (a) case1, (b) case2, (c) case3, (d) case4.

The four models are simulated in the CST Microwave Studio environment. The simulated reflection coefficients for these cases are shown in Figure V.11. From these curves, the performance of the antennas, in terms of resonance frequency and bandwidth spectrum, are extracted and listed in Table V.2. It can be seen from this table that the operating frequencies of the case 4 design are shifted towards the lower spectrum frequency, compared with the initial case 1, from 16.356 - 19.913 GHz to 5.134 - 5.684 GHz.

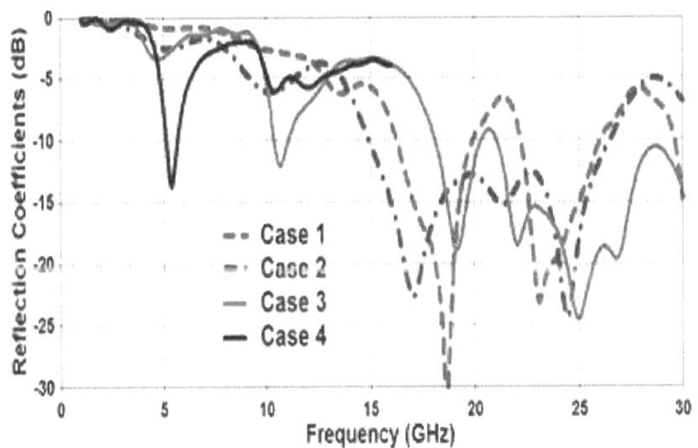

Figure V.11: Simulated reflection coefficient for the four cases.

Table V.2: Simulated results for the four antenna cases.

Antenna	$_{en}$ (GHz)	Bandwidth (GHz)
Case 1	18.632	16.356 - 19.913
Case 2	17.037	14.93 - 26.488
Case 3	10.657	10.374 - 11.056
Case 4	5.38	5.134 - 5.684

4.2 Antenna geometry

The optimum dimensions of the final antenna, shown in Figure V.12, are summarised in Table V.3.

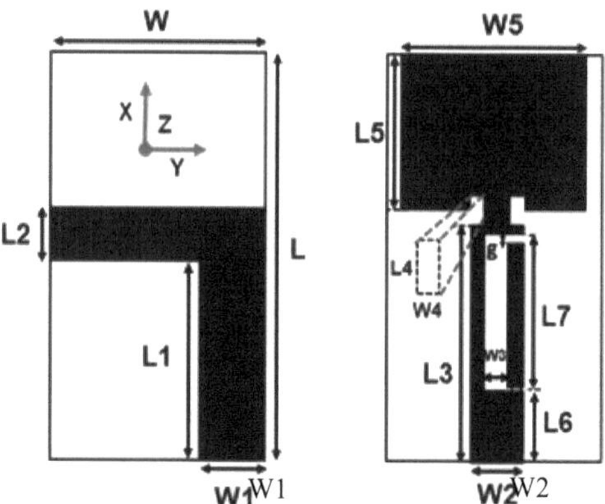

Figure V.12: Gëomëtrie of the final design.

Table V.3: Optimal proposëe antenna dimensions.

parameter	(mm)	parameter	(mm)	parameter	(mm)
W	8	W5	7	L4	1
W1	2.5	L	14.5	L5	5.5
W2	2	Li	7	L6	2.5
W3	0.8	L2	2	L7	5.5
W4	0.5	L3	8.5	g	0.1

4.3. Parametric study

To analyse the effect of substrate permittivity, excitation line dimensions and aperture (paramëtre g), ground plane gëomëtrie and patch gëomëtrie on antenna performance, a paramëtric ëtude is performed by the same simulation software in this section, acting on the paramëtres in Table V.3.

4.3.1. Effect of substrate permittivity

The effect of permittivity on the reflection coefficient is reprëzeйë in Figure V.13. From the curves in this figure, it can be clearly seen that the increase in substrate permittivity can dëcalate the resonant frequency of the microstrip antenna to the left (towards the lower frequencies).

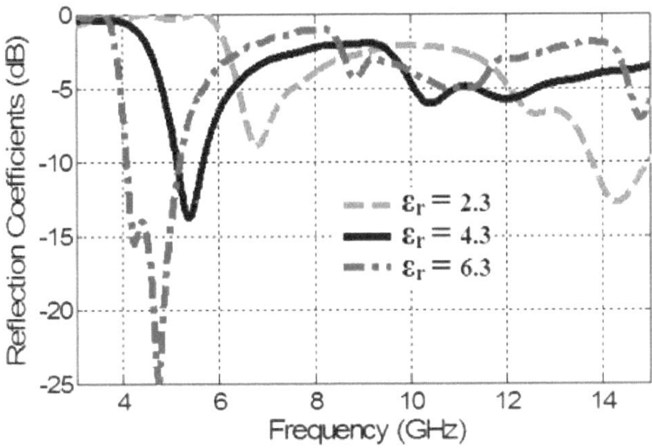

Figure V.13: Effect of substrate permittivity on the reflection coefficient.

4.3.2. Effect of the geometry of the excitation line

Figures V.14 and V.15 show the effect of the length (parameter L_3) and width (parameter W_2) of the microstrip feed line, respectively, on the reflection coefficient.

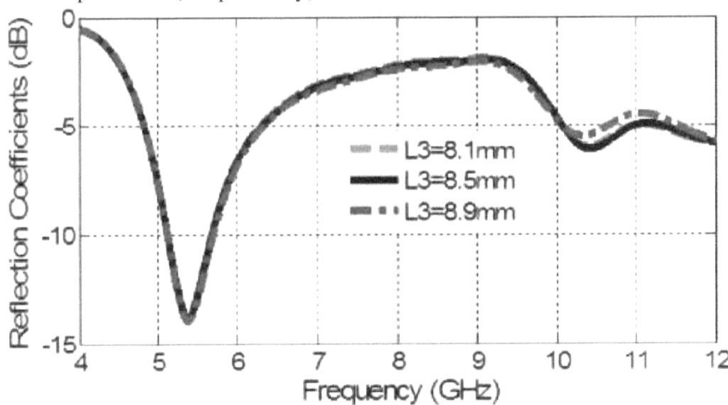

Figure V.14: Effect of parameter L3 on the reflection coefficient.

A small variation of 0.4 mm in the length of the excitation line does not change the resonance frequency. This is confirmed in Figure V.14, where the three curves are identical. This is in contrast to the line width, where there is a significant shift in the resonance frequency as a function of a 1 mm variation in the W2 parameter (Figure V.15).

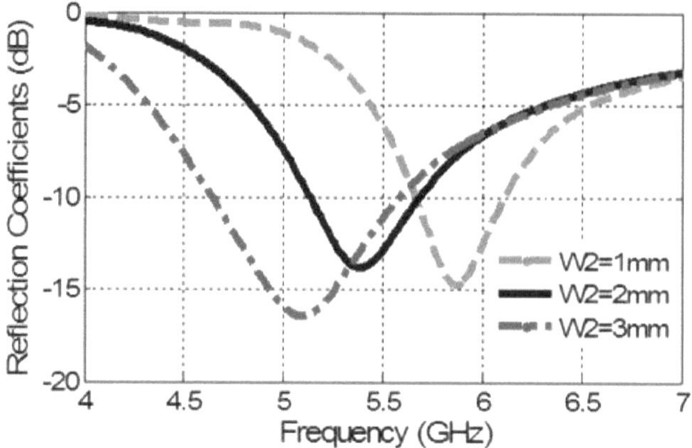

Figure V.15: Effect of parameter W2 on the reflection coefficient.

4.3.3. Effect of slot geometry in the excitation line

The effect of the aperture g -at the upper level of the feed line- on the reflection coefficient is illustrated in Figure V. 16. It can be seen that the optimum value which guarantees that the project covers the desired band (WLAN application) is g = 0.1 mm.

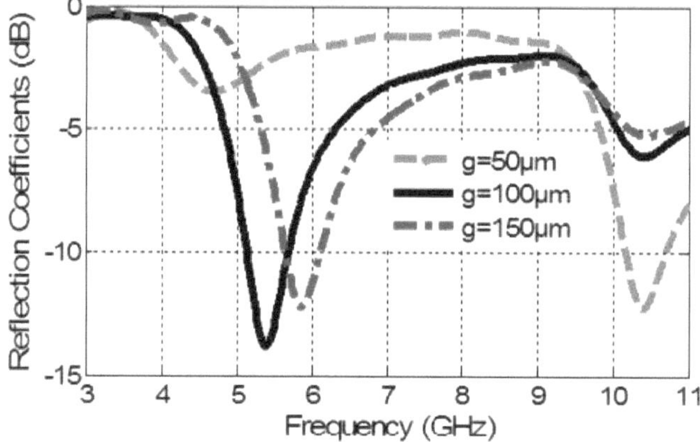

Figure V.16: Effect of parameter g on the reflection coefficient.

The effect of the gëomëtrie of the internal rectangular slot (the paramëtres L7 and w_3) engraved in the feed line- on the reflection coefficient is illustrated in Figures V.17 and V.18. From the curves in these figures, it can be seen that increasing the surface area of the rectangular slot (length or width) causes a degradation in the resonant frequency of the microstrip antenna.

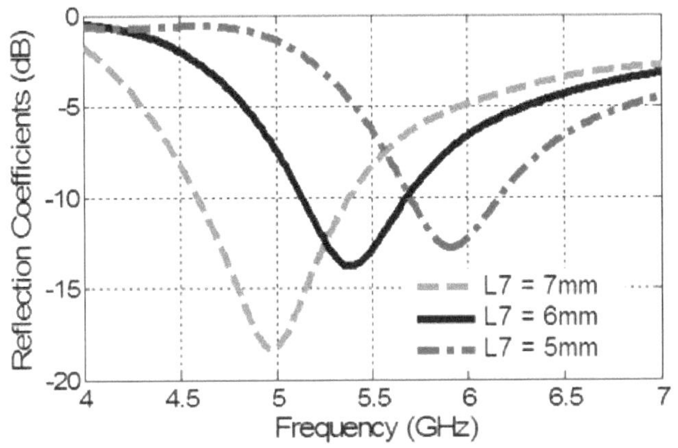

Figure V.17: Effect of parameter L6 on the reflection coefficient

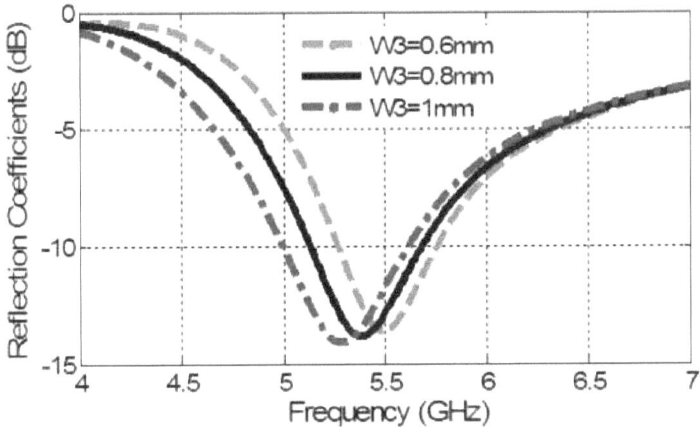

Figure V.18: Effect of parameter W3 on the reflection coefficient

4.3.4. Geometric effect of the radiating element (patch)

Figures V.19 and V.20 show the effect of the length (parameter L_5) and width (parameter W_5) of the radiating element (patch), respectively, on the reflection coefficient.

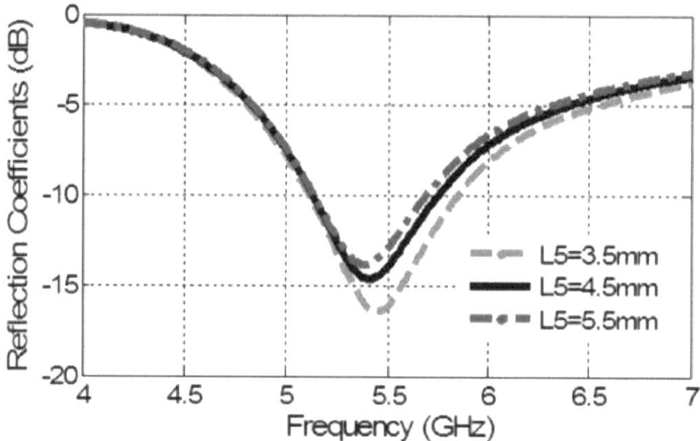

Figure V.19: Effect of parameter L5 on the reflection coefficient

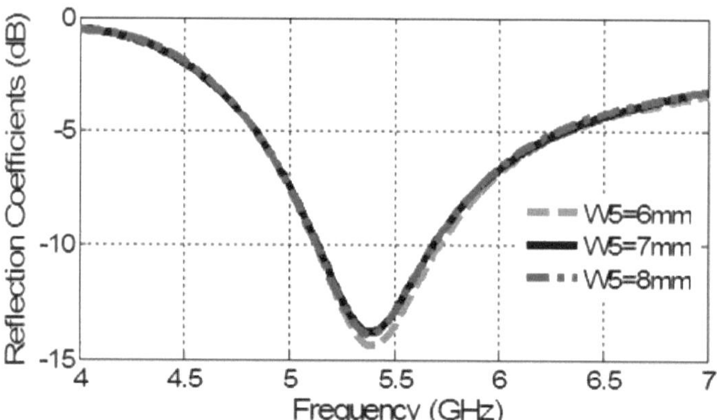

Figure V.20: Effect of parameter W5 on the reflection coefficient.

When we return to the curves in these figures, varying the length or width of the radiating element by 1 mm, we notice that the resonance frequency is not much changed and the curves in each figure are almost identical.

4.3.5. Ground plan geometry effect

The effect of the gc'ometry of the 'L-inverse' shape of the ground plane on the reflection coefficient is illustrated in Figures V.21 and V.22.

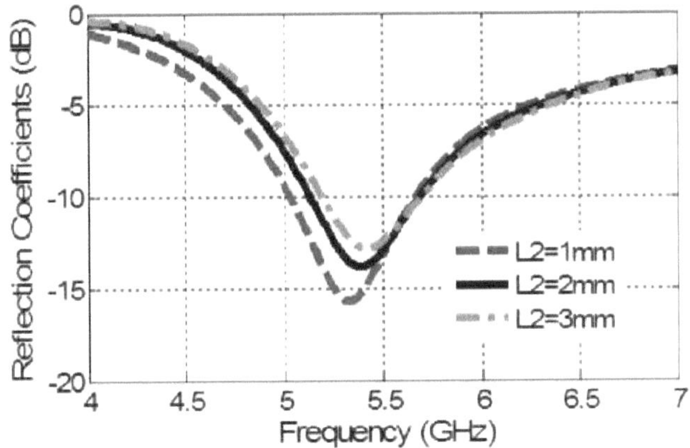

Figure V.21: Effect of parameter L2 on the reflection coefficient.

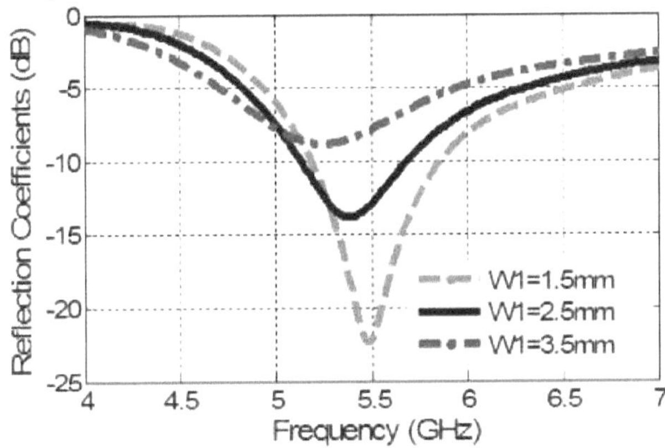

Figure V.22: Effect of w_i on the reflection coefficient.

From the curves in Figure V.21, it can be seen that increasing or decreasing the ground plane length (parameter l_2) by 1 mm, does not have a considerable effect on the reflection coefficient of the proposed antenna. On the other hand, in Figure V.22, the same variation by 1 mm in the ground plane width (parameter w_i) can modify the bandwidth of the antenna (matching) complementarily.

4.4. Results and discussion

In order to validate the numerical results, the final version of the antenna is manufactured and measured, as shown in Figure V.23.

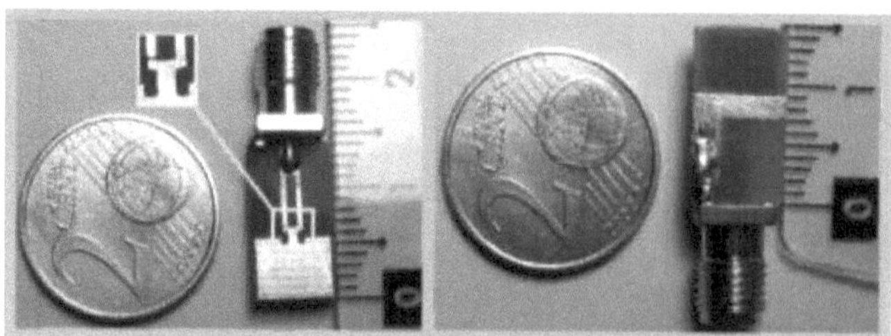

Figure V.23: Photograph of the prototype factory.

4.4.1. Reflection coefficient

The simulated and measured reflection coefficients of the proposed antenna are plotted together in Figure V.24. From this figure, it can be seen that the rectangular monopole antenna provides an impëdance bandwidth of 4.926 GHz and 5.198 GHz (s_{11} less than -10 dB) for the measurements and between 5.144 and 5.684 GHz for the simulations.

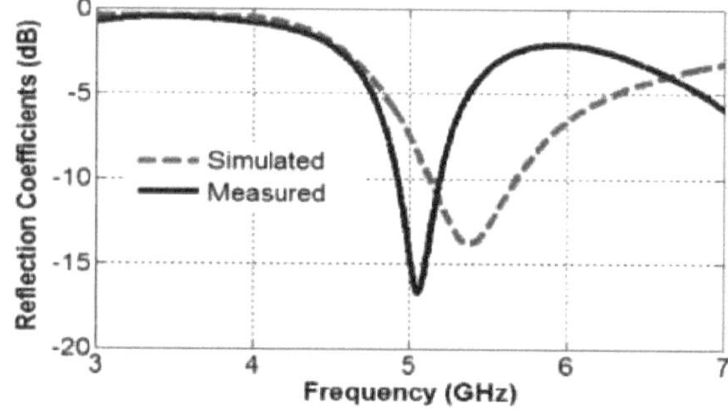

Figure V.24: Simulated and measured reflection coefficient of the proposëe antenna.

The discrepancy observed between the measured and simulated results is due to the effect of incorrect soldering of the SMA connector or connection errors on the SMA connector and the VNA cable (HF probe of the analyser). Manufacturing faults may also be the cause of this deviation in the measurement result.

Figure V.25 shows photos of paramëtre s_{11} measured by the N5224A vector network analyser.

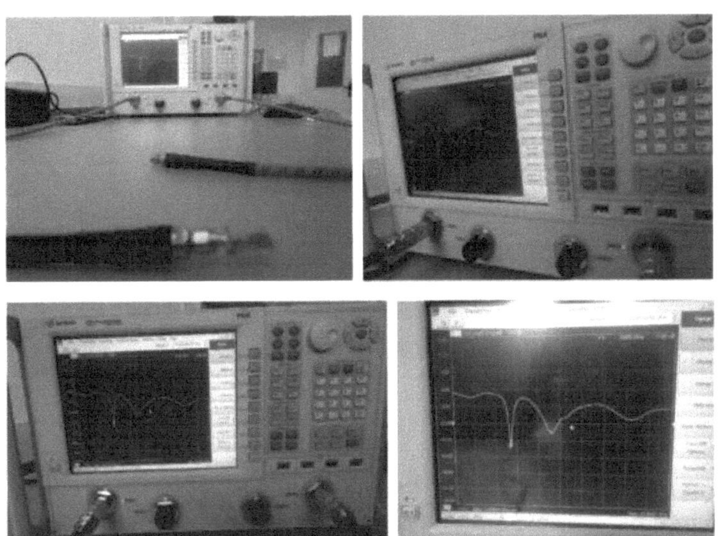

Figure V.25: Photographs of parameter s11 measured by the network analyser.

4.4.2. Distribution of magnetic field lines

Figure V.26 shows the field distribution of the antenna proposée at 5.38 GHz between the two conductors. At this frequency, the maximum field strength is concentrated on the right-hand side of the ground plane, its ëelectric length increases which explains the miniaturisation by the DGS technique.

Figure V.26: Field strength distribution of the proposed antenna simulated by the CST software
at 5.38 GHz.

4.4.3. Radiation diagram

The radiation patterns of the proposed design are calculated in the two principal planes (XZ and YZ planes) at a frequency of 5.38 GHz and shown in Figure V.27. The antenna radiation pattern is bidirectional in the XZ plane and omnidirectional in the YZ plane.

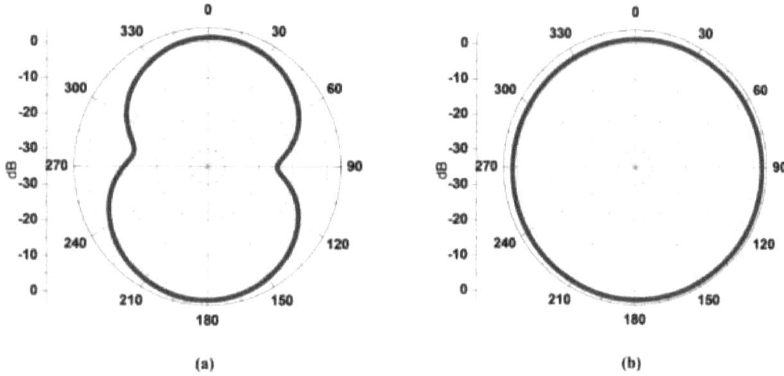

(a)   (b)

Figure V.27: Simultaneous radiation patterns at 5.38 GHz; (a) XZ plane, (b) plan YZ.

The simultaneous 3D radiation pattern at 5.38 GHz is shown in the figure. V.28.

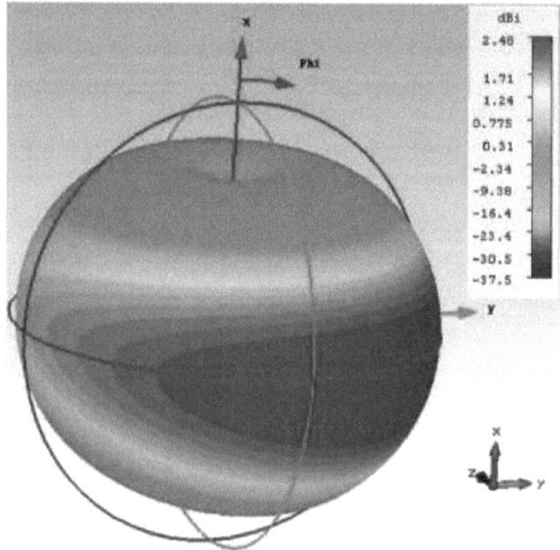

Figure V.28: The simultaneous 3D radiation pattern at 5.38 GHz.

4.4.4. Gain and efficiency

Figures V.29 and V.30 show, respectively, the actual gain and total efficiency of the proposed antenna simulated at 5.38 GHz. From the curves in these figures it can be concluded that the actual gain of the proposed antenna is about 2.1 dBi at 5.4 GHz. Furthermore, the efficiency is almost 84.67% at the resonant frequency.

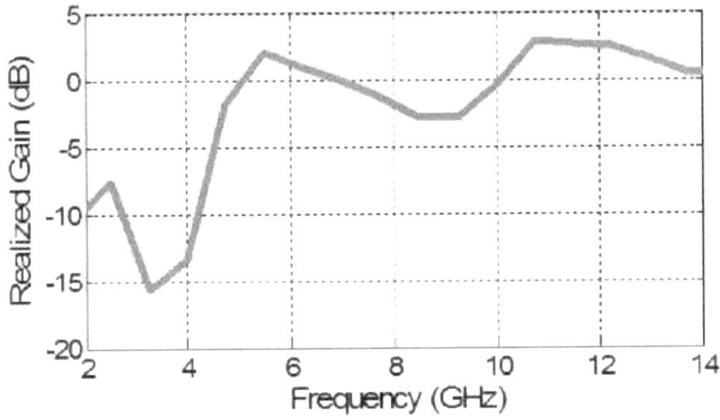

Figure V.29: Simulated antenna gain as a function of frequency.

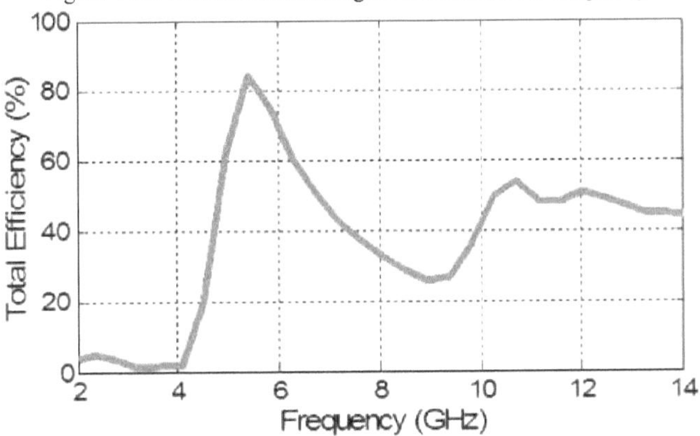

Figure V.30: Simulated total antenna efficiency as a function of frequency.

5. Wideband miniaturised monopole antenna [10]

In this work, a very small and compact printed rectangular monopole antenna is presented. The overall size of the antenna is 10 x 6 mm^2 and consists of a rectangular patch and an L-iinverse ground plane fed by a microstrip transmission line. The DGS structure technique is used to achieve miniaturisation.

5.1. Antenna design

The initial prototype consists of a rectangular patch antenna printed on a dielectric substrate with FR-4 (Fire Retardant) material with relative dielectric constant ;v = 4.4, ëthickness h =1.6 mm and loss tangent tanS = 0.02. The feed line and full ground plane are printed above and below the substrate, respectively, as shown in Fig. V.31.a. To achieve antenna miniaturisation, the same antenna structure is studied with a partially printed (L-inverted) ground plane, as shown in Fig. V.31.b.

The details of the two cases are as follows:

- **Case 1:** a rectangular patch antenna with a complete ground plane.
- **Case2:** a rectangular patch antenna with a DGS structure (inverse L-shaped).

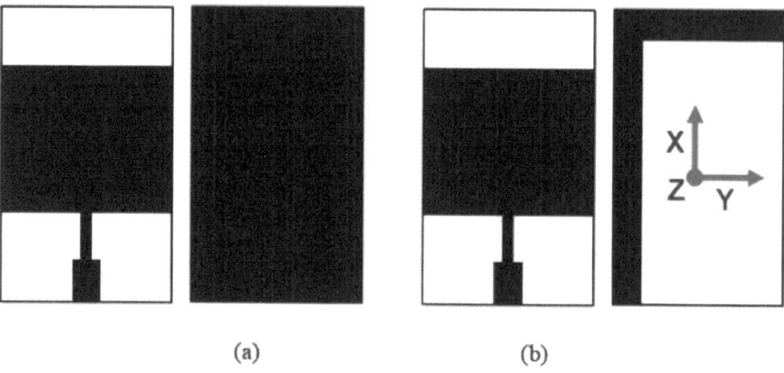

(a)   (b)

Figure V.31: Antenna design for (a) case1, (b) case2.

5.2. Theoretical considerations

To calculate the resonant frequency of the dominant mode analytically, the formula quoted by C. Balanis in [11] is applied:

$$f_r(\text{cas1}) = \frac{C}{2W\sqrt{\varepsilon_r}} \qquad (5.1)$$

Where; C is the celerity, W is the width of the radiating element, and ε_r is the relative permittivity of the substrate.

The frequency value obtained is approximately 12 GHz.

The structure in figure V.31.a is a classical patch antenna that resonates at half wavelength (X / 2), while the same antenna resonates at about a quarter wavelength (1 / 4) when the ground plane is reduced to an L-inverted shape (figure V.31.b). This explains why the resonance frequency in case 1 is almost double that obtained in case 2, as shown in Figure V.32.

For the modified ground plane monopole antenna, the resonant frequency can be given by the following expression [12]:

$$f_r(\text{cas2}) = \frac{C}{4W\sqrt{\varepsilon_r}} = \frac{f_r(\text{cas1})}{2} \qquad (5.2)$$

The new calculated value of the resonance frequency is closed at 5.96 GHz.

5.2. Numerical analysis

To compare the performance of different antenna cases, both designs were simulated using CST Studio software. All the electromagnetic characteristics (reflection coefficient, bandwidth, radiation pattern, gain and radiation efficiency) are analysed and discussed in this section.

5.2.1. Reflection coefficient

The reflection coefficient for the two cases is shown in Figure V.32.

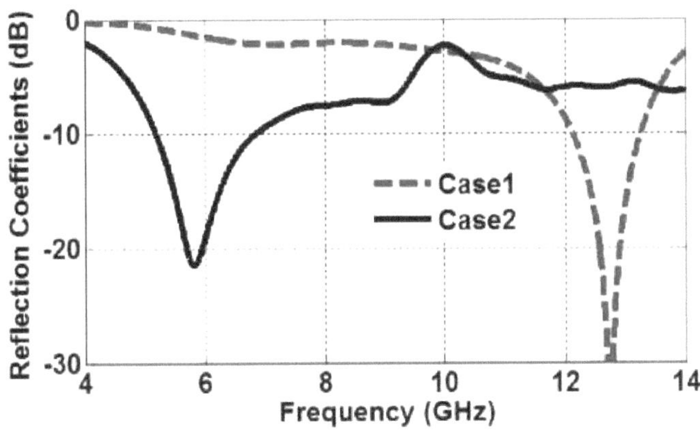

Figure V.32: Simulated reflection coefficients for the two cases.

From these curves, the performance of the antennas, in terms of resonance frequency and bandwidth spectrum, is extracted and listed in Table V.4. It can be seen from this table that the operating frequency of the Case 2 design is shifted towards the lower frequency compared with the initial design, from 12.1 - 13.3 GHz to 5.2 - 6.8 GHz. The miniaturisation factor is identified by 2.18. It is also noted in Table V.4 that the bandwidth is further increased from 1.12 GHz to 1.65 GHz with a difference of 531 MHz. This improvement is due to the effect of input impedance and antenna current flow.

Table V.4: Simulated results for two antenna cases.

Antenna	$_{en}$ (GHz)	Bandwidth (GHz)	BW %(GHz)	Gain/Total Efficiency (dBi/%)
Case 1	12.74	12.13 - 13.25	8.82(1.12)	4.58/70.62
Case 2	5.82	5.2 - 6.85	28.4(1.65)	2.34/97.55

5.2.2. Radiation diagram

The simulated radiation patterns of the two antennas are shown in Figure V.33 at their resonant frequencies (at 12.7 GHz for case 1 and at 5.8 GHz for case 2). From this figure, it can be seen that the radiation curve in case 2 is improved compared with the first case.

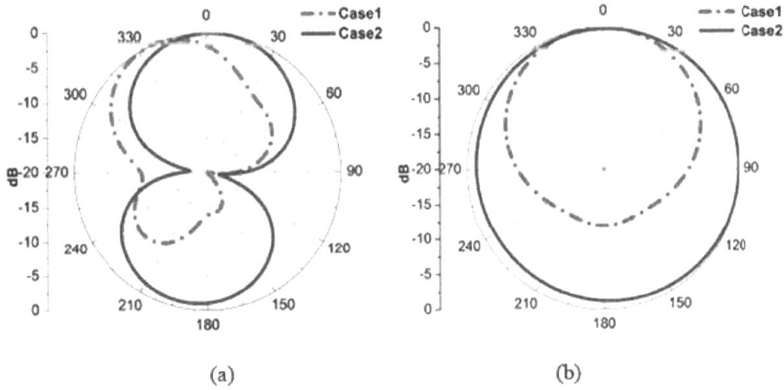

Figure V.33: Simulated radiation pattern for the two cases in;

(a) XZ plane, (b) YZ plane.

The simulated 3D radiation patterns of the two antennas are shown in Figure V.34 at their resonant frequencies.

In Figure V.34.a, the structure consists of a patch antenna where the ground plane is completely buried on the underside of the substrate. In this case, the patch radiates only in the +z direction caused by the reflection of electromagnetic waves on the ground plane. With the presence of the SMA connector, the radiation from the patch antenna is concentrated in its region. However, for the final design where the ground plane is partially removed, allowing radiation in both + z and -z directions (no reflection), it is clear that there is no such phënomëne of tilt (see Figure V.34.b).

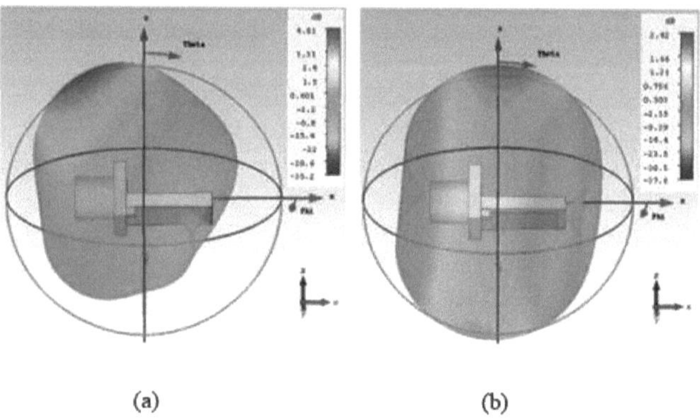

(a)   (b)

Figure V.34: Simulated 3D radiation pattern for; (a) casl, (b) cas2.

5.2.3. Gain and efficiency

Figures V.35 and V.36 show the simuted gain reяH3ë and total efficacile for the two cases, respectively.

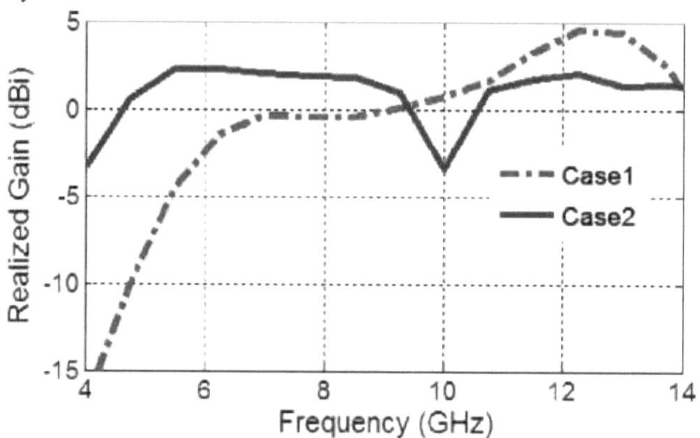

Figure V.35: Gain rëяH3ë simute for both cases.

From the curves in Figure V.35, it can be seen that the gain in case 2 is 2.34 dBi at 5.8 GHz and is almost halved compared with the initial case. According to Figure V.36, the maximum efficiency for case 1 is 70.62% at 12.5 GHz and 97.55% at 6 GHz for case 2. This increase can be explained by the configuration of the monopole antenna, where the diëlectrical loss of the substrate does not play a major role.

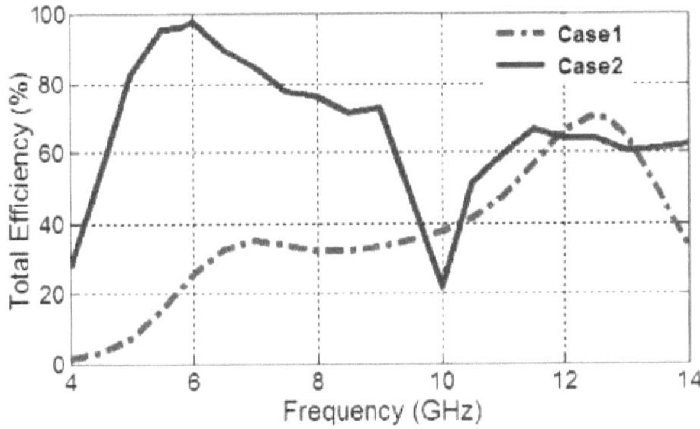

Figure V.36: Simulated efficiency for the two cases.

5.2.4. Field distribution

The magnetic field distribution of the original and new design between the two conductors is addressed and shown in Figure V.37 at their resonant frequencies.

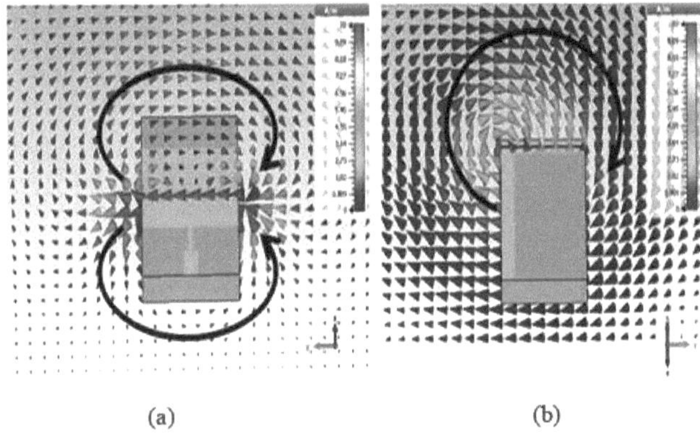

Figure V.37: Simulated magnetic field distribution for: (a) case 1 at 12.7 GHz, and (b) case 2 at 5,8 GHz.

The magnetic field distribution is homogeneous in Figure V.37.a, with the field lines emerging from one side of the substrate and moving towards the other end, propagating up and down the patch towards the global ground plane. On the other hand, for the partial ground plane (Figure V.37.b), the field lines take only one direction, with a long path, so the wavelength increases, which explains the miniaturisation by the DGS technique.

5.3. Antenna geometry

The optimum dimensions of the final antenna are shown in Figure V.38 and summarised in Table V.5.

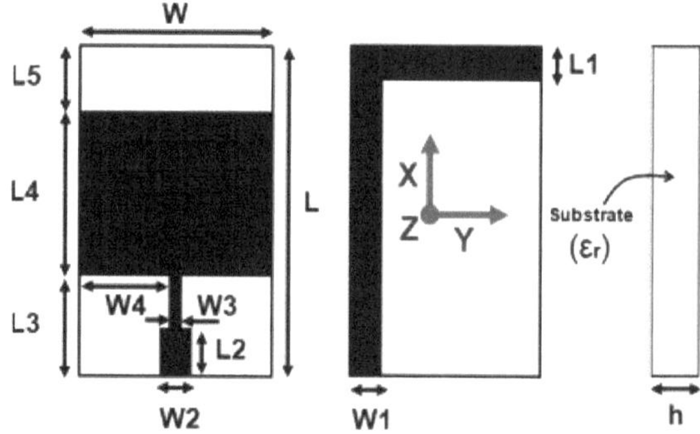

Figure V.38: Geometry of the final design.

Table V.5: Optimal dimensions of the proposed antenna.

parameter	value (mm)	parameter	value (mm)	parameter	value (mm)
L	10	Li	1	L2	1.5
L3	3	L4	5	L5	2
W	6	Wi	1	W2	1
W3	0.4	W4	2.8	h	1.6

5.4. Parametric study

Before manufacturing the prototype, a performance analysis of the proposed antenna was carried out to provide the best optimisation results. The same software is used to study the effect of permittivity and the effect of the geometry of the inverse L-shaped ground plane, the feed line and the geometry of the radiating element on the antenna's performance.

5.4.1. Effect of substrate permittivity

The effect of permittivity on the reflection coefficient of the proposée antenna is shown in Figure V.39. From the curves in this figure, it can be said that no shift in rysonance frequency is noticed with increasing substrate permittivity, but only the reflection coefficient value is modified (antenna matching).

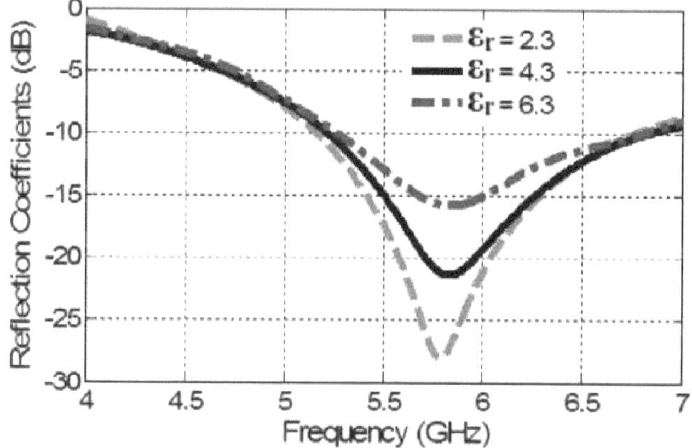

Figure V.39: Effect of substrate permittivity on the reflection coefficient.

5.4.2. Effect of the geometry of the excitation line

The effect of feedline gyometry (length L2 and width w_3) on the reflection coefficient is shown in Figures V.40 and V.41.

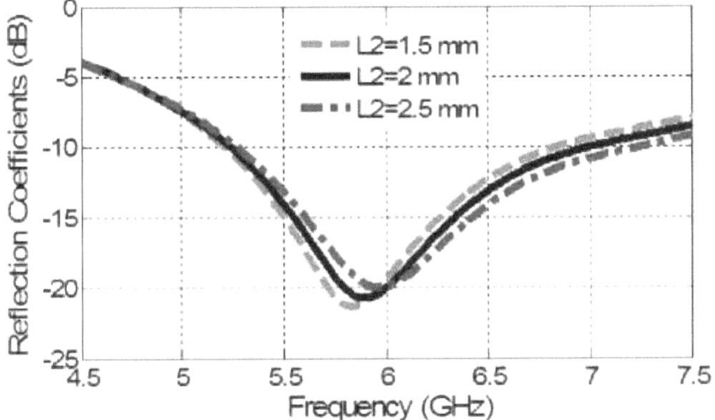

Figure V.40: Effect of parameter L2 on the reflection coefficient.

From the curves in these figures, it can be seen that the two parameters do not have a great effect on the reflection coefficient of the proposée antenna.

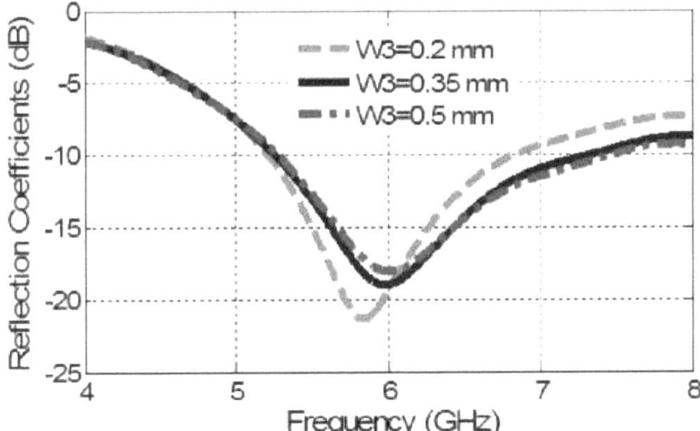

Figure V.41: Effect of parameter W3 on the reflection coefficient.

5.4.3. Geometric effect of the radiating element (patch)

Figures V.42 and V.43 show the effect of the length (parameter l_4) and width (parameter W) of the radiating element, respectively, on the reflection coefficient.

An increase of 2 mm in patch length can shift the resonance frequency to the right (see Figure V.42). The opposite is true for Figure V.43, where a 2 mm increase in the width W causes the resonance frequency to shift to the left.

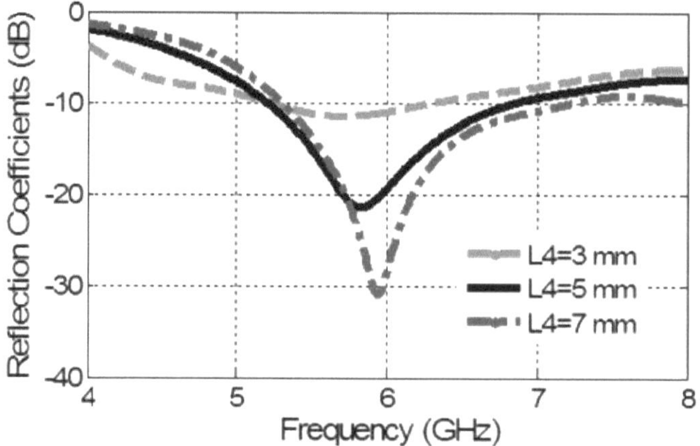

Figure V.42: Effect of parameter L4 on the reflection coefficient.

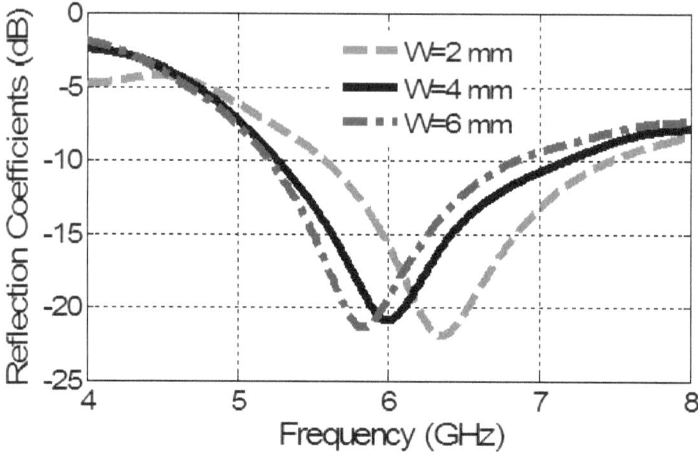

Figure V.43: Effect of parameter W on the reflection coefficient.

5.4.4. Ground plan geometry effect

Figures V.44 and V.45 show the effect of the length and width of the inverted I-shaped ground plane on the reflection coefficient. It is clear from these curves that to obtain a resonant frequency centred at 5.8 GHz (dedicated to the WLAN application), the length I1 = 1mm and the width W1 = 1mm should be chosen. We also note that the effect of the width W1 is greater than that of the length l_1, on the reflection coefficient of the proposed antenna.

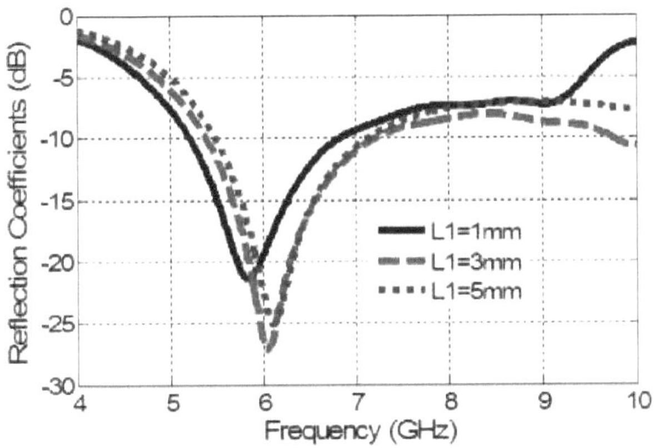

Figure V.44: Effect of parameter L₁ on the reflection coefficient.

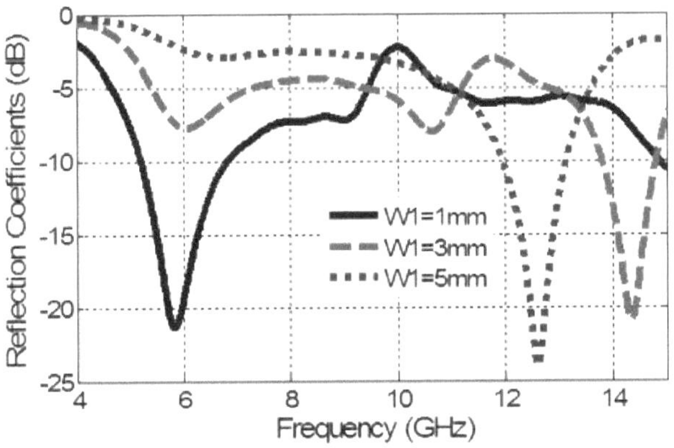

Figure V.45: Effect of w₁ on the reflection coefficient.

5.5. Measurement results and discussion

To analyse expërimentally the ëlectromagnëtic characteristics of the proposed antenna design, the final antenna of the antenna a ël.ë iabriquee (Figure V.46 shows the photograph of the fabricated prototype) and tested using an Agilent 8719ES vector network analyser.

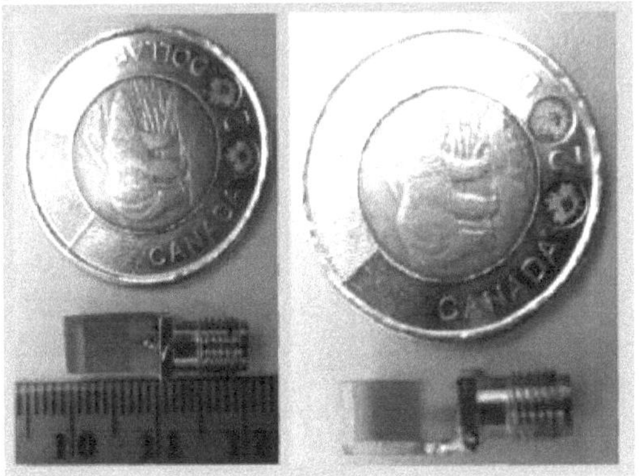

Figure V.46: Photograph of the prototype factory.

5.5.1. Reflection coefficient

The simulated and measured reflection coefficients of the proposed antenna are shown in Fig. V.47. It can be seen from this figure that the rectangular monopole antenna provides a wide bandwidth between 5.26 GHz and 6.28 GHz for the measured results (reflection coefficient less than -10 dB) and between 5.2 GHz and 6.85 GHz for the simulated results.

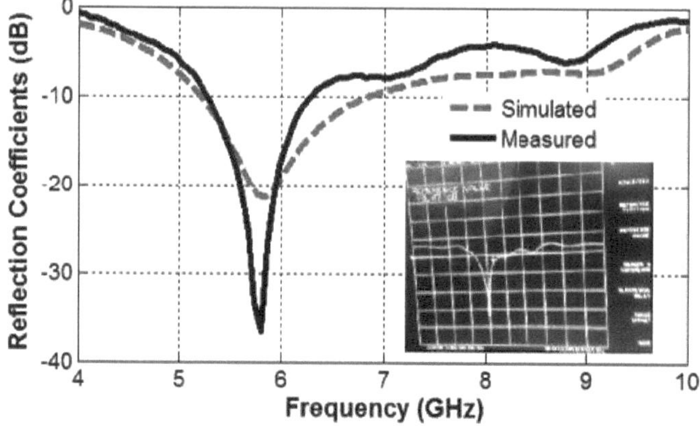

Figure V.47: Simulated reflection coefficient and measurement of the proposëe antenna.

Figure V.48 shows a photograph of parameter s_{11} measured by the Agilent 8719ES vector network analyser.

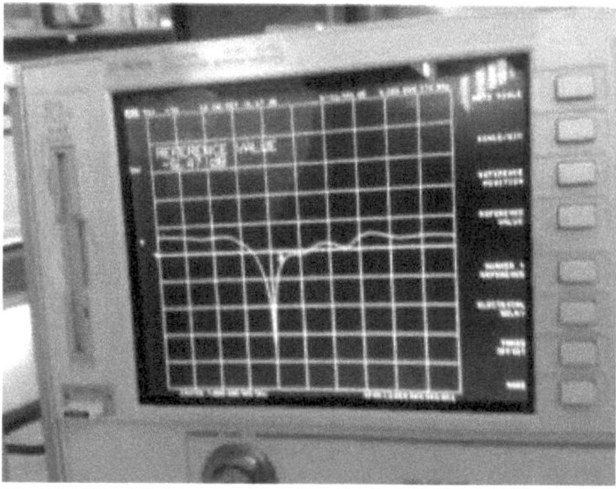

Figure V.48: Photograph of parameter s_{11} measured by the RV analyser.

5.5.2. Radiation diagram

The radiation patterns of the proposed design are calculated and measured in the two main planes (XZ plane and YZ plane) at the centre frequency of 5.8 GHz and representedësentës in Figure V.49. The radiation behaviour of the antenna is almost bidirectional in the XZ plane and omnidirectional in the YZ plane.

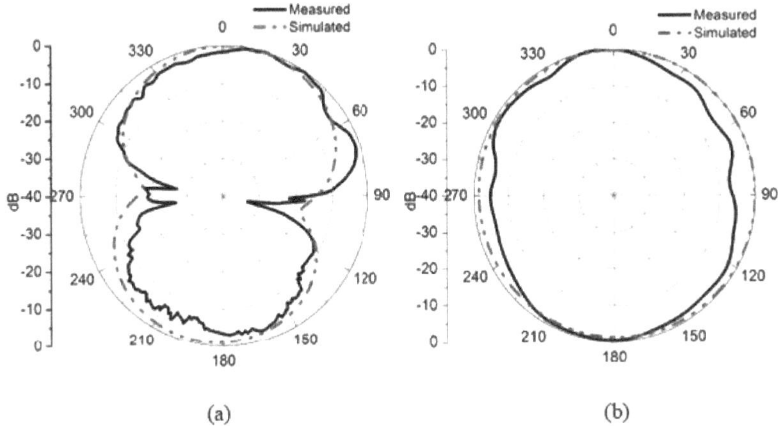

(a)   (b)

Figure V.49: Simulated radiation pattern and measurements at 5.8 GHz in; (a) XZ plane, (b) YZ plane.

5.5.3. Gain

Figure V.50 shows the gain геяНзё of the proposёe antenna obtained from simulation and measurements. From these curves, it can be seen that the gain геяНзё is almost stable in the operating frequency band and that the maximum gain value provided by the proposёe antenna is about 2.5 dB at the resonance frequency of 5.8 GHz.

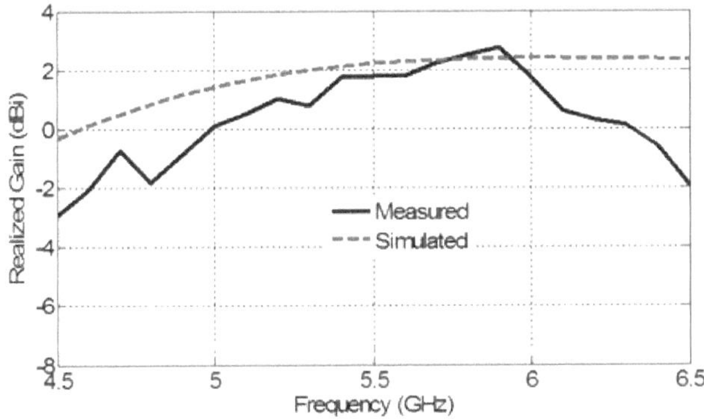

Figure V.50: Gain геяНзё simute and measurement of the proposёe antenna.

5.6. Comparison with recent antenna designs

In this section, the proposed antenna structure is compared (bandwidth and size) with recently published mono / dual-band antenna designs for the 5.2 / 5.8 GHz WLAN application (from the measured reflection coefficient less than - 10 dB).

Table V.6 shows a comparison between the proposed antenna and other recent antenna designs. From this table, it can be concluded that the proposed antenna is the most compact compared to the others and that it offers a bandwidth of 1 GHz that covers the WLAN band of the 802.11a standard.

Table V.6: Comparison between the proposed antenna and recent antenna designs.

Reference	en/Gain (GHz/dBi)	Bandwidth	Size including ground	Bandwidth at lowest

131

		(GHz)	plan (mm)^2	frequency
[13][a]	5.6/8.18	5.2-6.1	40 x 40	0.746X0X0.746X0
[14][a]	5.1/-	4.7-6.2	20 x 20	0.34X0X0.34X0
[15][b]	6.2/1.8	5 -7.2	15 x 15	O.31X0XO.31X0
[16][b]	5.2/8.2	5 - 5.2	36 x 30	O.624X0XO.52X0
[17][a]	5.8/14	5 - 6	23 x 23	0.444X0x0.444X0
[18][b]	5.2/1.71	5.1-5.4	24 x 16	0.416X0X0.277X0
[19][a]	5/6.5	4.8 - 6.6	n (32)?	0.945X0X0.945X0
[20][a]	5.8/5.1	5.1-6.83	11 x 13	0.213X0X0.251X0
[21][b]	6.3/<0	6.1-6.5	11 x 15	0.231X0X0.315X0
[22][a]	5.8/3.1	5.4-6	13 x 19.8	0.251X0X0.383X0
This work	5.8Z2.5	5.2-6.2	10 x 6	0.193X0X0.116X0

[a]) single band,[b]) dual or multi-band

6. Conclusion

In this chapter, we have presented and manufactured two printed rectangular monopole antennas (narrowband and broadband) miniaturised using the DGS technique:

For the first antenna discussed in Section 4, the use of the inverted L-shape in the ground plane and the slot in the feed line provides miniaturisation and reduces the resonant frequency by a factor of approximately 3. The result is a narrow bandwidth between 4.926 and 5.198 GHz for measurements and between 5.144 and 5.668 GHz for simulations (s_{11} less than -10 dB). The size of the proposed antenna is 14.5 x 8 mm^2 printed on a 1.6 mm thick FR-4 substrate.

For the second antenna presented in Section 5, the ground plane defect structure (GDS) technique is successfully used to reduce the ground plane by cutting a large slot to achieve miniaturisation (the etched rectangular slot represents a 75% reduction in the printed ground plane area which provides high miniaturisation). The structure of the ground plane consists of an inverted L-shape. The rectangular radiating element has a size of 6 x 5 mm2 and is connected to a microstrip supply line. The simulated and measured resonant frequency of the single-band antenna is about 5.8 GHz and can cover a bandwidth of more than 1 GHz for measurement and 1.65 GHz for simulation. The simulated and measured data are in good agreement. The proposed antenna is very compact (10 x 6 mm).^{2}

The bandwidth obtained for the two antennas covers the spectrum of the WLAN 802.11a band.

Bibliography of Chapter V

[1]   https://media.rs-online.com/t_large/F5265757-01.jpg

[2]   https://media.rs-online.com/t_large/F1939117-01.jpg

[3]   https://literature.cdn.keysight.com/litweb/pdf/5990-7745EN.pdf?id=2052675

[4]   http://vi.raptor.ebaydesc.com/ws/eBayISAPI.dll?ViewItemDescV4&item=14239619
8652 &category=40004&pm=1&ds=0&t=1516404847000&ver=0

[5]   https://www.keysight.com/en/pdx-x201876-pn-N5224A/pna-microwave-network-analyzer-435-ghz?pm=spc&nid=-32497.1150216&cc=DZ&lc=eng

[6]   https://www.testworld.com/wp-content/uploads/user-guide-keysight-agilent-hp-8719et-
8720et-8722et-8719es-8720es-8722es-network-analyzers_new.pdf

[7]   These PhD in дёше électroшque and électrics, Shaozhen Zhu, 'Wearable Antennas For Personal Wireless Networks', Universe of Sheffield, January 2008.

[8]   http://www.s3m-faraday.fr/mediastore/1/35107_1_FR_110_x.jpg

[9]   Farouk Chetouah, Nacerdine Bouzit, Idris Messaoudene, Salih Aidel, Massinissa Belazzoug, Youcef Braham Chaouche, 'Miniaturized printed rectangular monopole antenna with a new DGS for WLAN applications', International Symposium on Networks, Computers and Communications (ISNCC2017), Pp. 1-4, Marrakech, Morocco, 16-18 May 2017.

[10]   Farouk Chetouah, Salih Aidel, Nacerdine Bouzit, Idris Messaoudene, 'A miniaturized printed monopole antenna for 5.2-5.8 GHz WLAN applications', International Journal of RF and Microwave Computer-Aided Engineering, 2018.

[11]   antenna Theory: Analysis Design, Third Edition, Constantine A. Balanis, Chapter 14: Microstrip Antennas, Pp. 826-831, John Wiley & Sons, Inc, 2005.

[12]   J. Yang, H. Wang, Z. Lv, and H. Wang, "Design of Miniaturized Dual-Band Microstrip Antenna for WLAN Application," Sensors, Vol. 16, No. 7, Jul. 2016.

[13]   Elsdon M, Yurduseven O, Dai X. *Wideband metamaterial solar cell antenna for 5 GHz Wi-Fi communication.* Prog Electromagn Res C. 2017 ;71:123 - 131. Vol.

[14]   Ojaroudi Y, Ojaroudi N, Ghadimi N. *Circulary polarized microstrip slot antenna with a pair of spur-shaped slits for WLAN applications.* Microw Opt Technol Lett. 2015 ;57(3):756 - 759.

[15]   Alibakhshi Konari M. *Miniaturized printed monopole antenna with applying the modified conductor-backed plane and three embedded strips based on CPW for multi-band telecommunication devices.* Int J Microw Wireless Technol. 2016 ;8(08):1 - 5

[16]   Yang H-L, Yao W, Yi Y, Huang X, Wu S, Xiao B. *A dual-band low-profile metasurface-enabled wearable antenna for WLAN devices.* Prog Electromagn Res C. 2016 ;61:115 - 125.

[17]   Liu L, Li Y, Zhang Z, Feng Z. *Compact helical antenna with small ground fed by spiralshaped microstrip line.* Electron Lett. 2014 ;50(5):336 - 338.

[18]   Chakraborty U, Kundu A, Chowdhury SK, Bhattacharjee AK. *Compact dual-band microstrip antenna for IEEE 802.11a WLAN Application.* IEEE Antenna Wireless Propag Lett. 2014 ;13:407 - 410.

[19]   Wong H, So KK, Gao X. *Bandwidth enhancement of a monopolar patch antenna with V- shaped slot for car-to-car and WLAN communications.* IEEE Trans Vehic Technol. Mar. 2016;65(3):1130 - 1136.

[20] Liu W-C, Hu Z-K. *Broadband CPW-fed folded-slot monopole antenna for 5.8 GHz RFID application.* Electron Lett. 2005 ;41 (17):937 - 939.

[21] Alam MJ, Faruque MRI, Islam MT. *Split quadrilateral multiband microstrip patch antenna design for modern communication system.* Microw Opt Technol Lett. 2017 ;59(7):1530 - 1538.

[22] Pandeeswari R, Raghavan S. *Meandered CPW-fed hexagonal split-ring resonator monopole antenna for 5.8 GHz RFID applications.* Microw Opt Technol Lett. 2015 ;57(3):681 - 684... No.

General conclusion

This book presents the analysis, design and modelling of new planar microwave structures for communication system applications. Our motivation has ël.ë the dëfi of improving the bandwidth of small, low-cost antennas that are easy to manufacture.

In this work, the ëlectrical and ëlectromagnëtic characteristics of planar antennas have ël ë discutëes by comparing the two technologies; microstrip and diëlectrical resonator.

Several tlK'oric approaches for antennas (ESA) with different miniaturisation techniques and the most relevant research work have been described.

A new miniaturised dielectric resonator antenna has been numerically studied and analysed using two electromagnetic simulators: Ansoft HFSS and CST. The structure studied consists of a rectangular dielectric resonator stacked with a thick layer of dielectric material that has a very high permittivity. Using this technique, the radiating element can be reduced by up to 90%.

The BST ceramic material with high dielectric permittivity is used to achieve this miniaturisation. Their dielectric properties (field-dependent dielectric constant, tunability and loss tangent), and the techniques for developing BST thin films are presented.

The results obtained are expressed in terms of the reflection coefficients and radiation characteristics of the proposed antenna with the effects of the dielectric resonator parameters on the reflection coefficient. The proposed antenna operates around 2.31 GHz and provides a 20 MHz impedance bandwidth between 2.31 GHz and 2.33 GHz. With these characteristics, the proposed RD antenna may be a suitable candidate for Wireless Communications Service (WCS) systems.

The description of the measurement equipment used and the manufacturing procedure for the prototype antennas were discussed. A numerical and experimental study and analysis of two microstrip antennas connected to a miniaturised microstrip feedline using the DGS 'inverse L' technique was carried out. The advantages of these fabricated antennas are: simple design with low-cost material, miniaturised patch size and 802.11a WLAN band spectrum.

The results show that the second antenna, which has a smaller patch area and an overall size of 10×6 mm^2, has a wide bandwidth of more than 1 GHz. In contrast, the first antenna can cover a narrow spectrum of 272 MHz, despite its larger overall size (14.5×8 mm2). This is explained by the effect of the size of the ground plane and its location (position of the 'inverse L' shape) on the miniaturisation and bandwidth of the microstrip antenna.

The three projects discussed clearly show the importance of the technique of loading a rectangular diëlectrical resonator antenna using a thin film of very high permittivity material and the technique of ground plane defect structure (GDS) in the miniaturisation of planar antennas.

Finally, our research work is of definite interest in the field of miniaturisation of planar antennas, and will undoubtedly enrich wireless applications, particularly WLANs. It may be extended by work with the following perspectives:

- Design of planar antennas using other miniaturisation techniques.
- These techniques can be applied to multi-port antennas (MIMO antennas).
- Improving the gain of electrically small antennas (ESA).
-

Printed by Books on Demand GmbH, Norderstedt / Germany